Every Home a Distillery

EARLY AMERICA: HISTORY, CONTEXT, CULTURE

Joyce E. Chaplin and Philip D. Morgan, Series Editors

Every Home a Distillery

Alcohol, Gender, and Technology in the Colonial Chesapeake

SARAH HAND MEACHAM

The Johns Hopkins University Press
Baltimore

© 2009 The Johns Hopkins University Press
All rights reserved. Published 2009
Printed in the United States of America on acid-free paper
2 4 6 8 9 7 5 3 1

The Johns Hopkins University Press
2715 North Charles Street
Baltimore, Maryland 21218-4363
www.press.jhu.edu

Library of Congress Cataloging-in-Publication Data
Meacham, Sarah Hand, 1972–
Every home a distillery: Alcohol, gender, and technology in the colonial
Chesapeake / Sarah Hand Meacham.
p. cm.
Includes bibliographical references and index.
ISBN-13: 978-0-8018-9312-4 (hardcover : alk. paper)
ISBN-10: 0-8018-9312-7 (hardcover : alk. paper)
1. Brewing — Social aspects — Chesapeake Bay Region (Md. and Va.) —
History. 2. Distilling industries — Social aspects — Chesapeake Bay
Region (Md. and Va.) — History. 3. Housewives — Chesapeake Bay
Region (Md. and Va.) — History. 4. Home economics — Chesapeake Bay
Region (Md. and Va.) — History. 5. Sex role — Chesapeake Bay Region
(Md. and Va.) — History. 6. Social classes — Chesapeake Bay Region
(Md. and Va.) — History. 7. Drinking of alcoholic beverages —
Chesapeake Bay Region (Md. and Va.) — History. 8. Bars (Drinking
establishments) — Chesapeake Bay Region (Md. and Va.) — History.
9. Chesapeake Bay Region (Md. and Va.) — Social life and
customs — 17th century. 10. Chesapeake Bay Region
(Md. and Va.) — Social life and customs —
18th century. I. Title.
TP573.U6M43 2009
641.2′1097409033 — dc22   2008050544

A catalog record for this book is available from the British Library.

*Special discounts are available for bulk purchases of this book. For more
information, please contact Special Sales at 410-516-6936
or specialsales@press.jhu.edu.*

The Johns Hopkins University Press uses environmentally friendly book
materials, including recycled text paper that is composed of at least
30 percent post-consumer waste, whenever possible. All of our
book papers are acid-free, and our jackets and covers
are printed on paper with recycled content.

For Scott and Camden

Contents

Preface

This project began when I was trying to research an early-eighteenth-century tavern known as Susannah Allen's in colonial Williamsburg. It has long been rumored that Allen's tavern was a brothel. While I suspected that there were not enough single women in colonial Virginia to have supported a brothel, I was intrigued. Very few records of Allen's tavern remain, and I was unable to determine whether her tavern sold sex as well as drinks and lodging. In the course of my research, though, I began to wonder whether Susannah Allen was unusual, or whether other taverns were run by women. The answers I discovered ultimately became chapter 4 of this book.

During that initial investigation, I ran across a small notation indicating that Susannah Allen had purchased some cider for her tavern from a woman. I wondered if the note was mistaken—had someone written "Mrs." for "Mr."? Was a planter's wife handling the plantation's affairs while her husband was out of town? This book is my answer to that question.

Every Home a Distillery covers mostly English experiences in the Chesapeake from 1690 to 1800. The book begins in 1690 because that is the earliest date from which sufficient records are available. There simply is not enough material to write a well-researched account of alcoholic beverage production in the earliest years of English settlement, in part because the Native Americans in the area, the Powhatan, are not known to have made alcoholic beverages.

The book ends in 1800 because one of its goals is to research the lives of colonial women in the South, and there are already books about early republic and antebellum women for those who wish to explore later years. As history would have it, women generally tend to appear only through careful readings of probate records, wills, court documents, and account books, all written by men. During this period most Chesapeake women of all races were illiterate, and no diaries or letters from colonial Chesapeake women are known to remain today. The book ends in 1800 also because massive German immigration in the 1840s changed

America's drinking dramatically. The Germans brought lager beer to the United States and built large breweries, both of which have been studied considerably.

Finally, although the history of alcoholic beverage production suggested the promise of detailing the meeting of three worlds in the Chesapeake (Native American, African, and English), a lack of manuscripts has reduced this study to mostly English America. Although references are made to the enslaved community when possible, the extant records of how people made alcohol in the colonial Chesapeake simply do not allow for more cross-cultural analysis.

While the frustrations of this study may now be evident, its fascinations might require explanation. No one has yet studied the production of alcoholic beverages in colonial America. Discovering and demonstrating that making alcohol used to be women's work and that taverns were passed down through women has been exciting. Furthermore, alcohol records have revealed how colonial communities worked together to produce, distribute, and consume the alcoholic beverages that they desperately needed. Each household played a role in a series of economic transactions, the scope of which the colonists probably were unaware. The history of making alcohol in the Chesapeake also reveals that colonists used innovations in technology to become more self-sufficient and insular over time. Finally, alcohol records provide a glimpse of a world that is lost and is no longer fully understandable — a world in which alcohol was critical to survival. The shift to a world where drinking is highly circumscribed began in the late eighteenth century.

It is my great pleasure to thank the organizations that have supported this project. Virginia Commonwealth University provided a College of Humanities and Sciences Career Enhancement Scholarship that allowed me to spend a summer writing. Fellowships from the American Philosophical Society; the David Library of the American Revolution; the Dibner Library at the National Museum of American History; the Early American Industries Association; the Jamestowne Society; the Program of Early American Economy and Society at the Library Company of Philadelphia; the Rockefeller Library at the Colonial Williamsburg Foundation; the Virginia Historical Society; the Winterthur Library, Museums, and Gardens; and an Albert S. Beveridge grant from the American Historical Association all supported the archival research necessary for this book.

This project began as a dissertation under the direction of Dr. Peter Onuf at the University of Virginia. Peter has always been extraordinarily generous with his time and suggestions. It has been my privilege to be his student.

My thanks to *The Virginia Magazine of History and Biography*, which published chapter 2 in an earlier version as " 'They Will Be Adjudged by Their Drinke What Kind of Housewives They Are': Gender, Technology, and Household Cidering in England and the Chesapeake, 1690 to 1760" and to the anonymous readers of that submission who offered such helpful guidance. Likewise, the reports of anonymous readers for *Early American Studies*, which published parts of chapter 4 as "Keeping the Trade: The Persistence of Tavernkeeping among Middling Women in Colonial Virginia," strengthened that chapter and have my gratitude.

Finally, editor Robert J. Brugger and the anonymous readers who critiqued the book manuscript for the Johns Hopkins University Press have offered the most useful advice in the gentlest terms. Their support of a first-time author is profoundly appreciated.

This book is dedicated to my husband and our daughter.

Every Home a Distillery

Introduction

"Do you know Jack?" begins some Jack Daniel's distillery websites and brochures. Readers learn that "Mr. Jack" bought his first still at the age of thirteen in 1863 and founded the "nation's oldest registered distillery" three years later. Readers are told that "Mr. Jack" always dressed in "a formal knee-length frock coat and a broad-brimmed planter's hat." Thus America's largest distilling company suggests that its whiskey is really the offering of an elite southern planter. Consumers who drink Jack Daniel's whiskey are invited to fantasize that, by imbibing, they too become grand southern planters with the wealth and leisure required for the home distillation of whiskey and the appreciation of it. In this fantasy there are no inconvenient lower-class men or women or enslaved laborers — only a wealthy planter who has made some whiskey and shares it with his equally wealthy guests as a sign of ease, generosity, and intimacy.[1]

In reality, large planters were unlikely to distill their own alcoholic beverages by the 1860s, and they were very unlikely to share them for free. The whiskey that they drank, and the enslaved men who made it, were recent interlopers in the history of alcoholic beverage production. Distilling remained a rather new and still-uncertain technological process and had but a short while earlier been almost exclusively the work of women.

This book investigates the world of alcoholic beverage production from the late seventeenth century to the late eighteenth century in the Chesapeake, the increasingly settled portions of Virginia and Maryland that touched the Chesapeake Bay and its rivers. Alcoholic drink was one of the few items that colonists could not live without. In a place where the water was unsafe, milk was generally unavailable, tea and coffee were too expensive for all but the very wealthy, and soda and nonalcoholic fruit juice were not yet invented, alcoholic beverages were all that colonists could drink safely. Even so, colonists imbibed startling amounts of alcohol. By 1770, the average adult white man drank the equivalent of seven shots of rum per day, and an average white woman drank almost two pints of hard cider per day. Children consumed alcoholic drinks daily, as slaves likely did as well. White servants were guaranteed alcoholic beverages by their employment contracts. Apprentices took daily breaks for drinks with their masters. Men running for office wooed voters with alcoholic concoctions. Women not only consumed alcoholic drinks, but they also cleaned houses and babies with them and used them as beauty products. Men and women, white and black, used alcoholic mixtures as medicine in attempts to cure headaches, lockjaw, melancholy, and colds.

Colonists drank alcoholic beverages during church services, in courtroom proceedings, in the House of Burgesses, at quiltings, at home, at taverns, at weddings, and at funerals. They drank raw ciders of apples, peaches, persimmons, and other fruits; they drank ciders that had been distilled into apple and peach brandies and cherry "bounce"; they drank molasses made into rum, and, occasionally, a little wheat made into ale. And toward the end of the eighteenth century, they turned corn into whiskey.

And yet, despite the obvious importance of alcohol to Chesapeake colonists, historians do not know how they acquired all these drinks. The significance of alcoholic beverage *consumption* in the American colonies emerges from several excellent books. One scholar has demonstrated that Massachusetts colonists used taverns to contest, first the authority of the Puritan elite, and then the authority of the English, helping to foment the American Revolution. Another found that tavern-going helped spread revolutionary ideas in Philadelphia as well, although with the ironic twist that, while republican men spouted notions of equality, the taverns they patronized were increasingly stratified by class. A third study established that Europeans courted trade with Native Americans by selling them alcohol, which Indians then adopted into their mourning and other rituals. Finally, a fourth scholar focused on the temperance movement from the 1780s to the 1830s and determined that the movement began as a reaction against cheap corn whiskey made by western settlers. All of these works offer nuanced and persuasive accounts of alcoholic beverage *consumption* in early America, but *Every Home a Distillery* is the first book to examine how colonists actually *acquired* their alcohol in the eighteenth century.[2]

Just how colonists came by their drink turns out to be a complex and revealing story. The limited materials related to alcoholic beverage production that have survived to this day — a few planter and tavern account books, a handful of cookbooks, several works of farm management advice literature, wills, probate records, advertisements, court records, and four planter "journals" — reveal that the Chesapeake was far more local and provincial than its elites would have admitted.

These records divulge that Chesapeake colonists lagged behind Europe, New England, and the Middle Colonies, in terms both of who was making alcoholic drinks and of how they were making them. Chesapeake colonists continued to rely on women to make alcohol for at least a hundred years after Europe, New England, and the Middle Colonies had turned to alcoholic beverages produced by men. Chesapeake women were still laboring over raw fruit alcohols such as cider at a time when men in the rest of the English empire produced hopped and distilled liquors.

In addition, the Chesapeake records disclose that small planters depended on large planters for alcohol in the early eighteenth century, since large planters were the only ones who could afford the stills that allowed ciders to be turned into nonspoiling brandies. As soon as smaller, cheaper stills became available in the Chesapeake in the second half of the eighteenth century, small planters leapt at the chance to make their own alcoholic drinks independently and year-round. Whenever they could, all planters tended to make their own alcoholic drinks rather than rely on imports. The alcohol that Chesapeake colonists made was peculiar to their region: they brewed and distilled with ingredients such as persimmons that were unfamiliar to Europe and New England. In many ways, Chesapeake colonists became increasingly insular and local over time. Finally, the records indicate that drinking became a matter of social concern in the Chesapeake in the late eighteenth century in ways that were very different from those of other regions of America.

This examination of how colonists got their alcohol begins by explaining the varieties of drinks and drinking places and the reasons that colonists drank so much. Simply put, Chesapeake colonists drank because they came from a tradition of heavy drinking and because they had little to drink that was not alcoholic. White colonists usually drank in same-sex groups, with men drinking out-of-doors and women indoors. Meals and funerals, where men and women drank together, were exceptions. Chesapeake colonists also needed alcoholic beverages to control pain, to perform beauty regimens, to clean, and generally to foster a sense of community.

The early Chesapeake was anomalous in the Atlantic world. In most of the world, including Western Europe, Latin America, and New England, the introduction of new technologies, particularly the flower of the hop plant, had led men to assume control of alcoholic beverage production by or during the seventeenth century. In contrast, seventeenth-century Chesapeake men, fixated on their tobacco monoculture, continued to rely on cider produced by women. Unlike women in other countries such as England, Holland, and Peru, Chesapeake women did not enjoy any particular economic control or status because of the alcohol they produced.[3]

In the first half of the eighteenth century, women in small-planter households typically produced cider only from July to December. These households lacked the technology to preserve the cider they produced, and when they ran out, they had a unique problem: unlike people elsewhere, Chesapeake colonists could not purchase alcoholic beverages at a local market or a commercial distillery or brewery. Most colonists could not import foodstuffs because they had neither the connections nor the financial credit and reputation in England or the West Indies to do so. The irregularity of shipments and the difficulties inherent in shipping beverages made it impractical for even wealthier Chesapeake colonists to import much liquor. So when small planters of the early eighteenth century ran out, they turned to large planters for their alcohol, at a price. Only large planter households had the technology — the cider presses, barrels, cellars, bottles, stills, and grafting talent — that allowed them to make alcoholic beverages on a year-round basis. A large planter could earn six times the average annual income of a small planter or laborer through selling surplus cider alone.

Taverns were not an alternative to large planters as a place to purchase alcoholic beverages; they were often just extensions of the large plantations themselves. Wealthy planters determined who could run a tavern and sold tavernkeepers foodstuffs that they had produced or imported. Court records show that only those small planters who were favored by their larger counterparts received tavern licenses. Interestingly, large planters and courts typically granted tavern licenses to middling men whose wives had previous tavernkeeping experience, a fact that scholars have not realized. Tavernkeeping was not a source of power or prestige for colonial women in the early Chesapeake. Rather, keeping a tavern, like the task of making alcoholic beverages, was something that women did as part of their share of the household labor.[4]

Small-planter households resented their dependence on large-planter households. Although the Chesapeake continued to lag behind Europe, the arrival during the second half of the eighteenth century of the three-gallon alembic still, a series of improved cider presses, the newly developed Hewes crab apple, and other technologies allowed small-planter households to become more self-sufficient. They developed alcohol trade networks with kin and people of their own kind. After 1760, the small stills, technical books, and knowledgeable new scientific and agricultural societies helped make *Every Man His Own Distiller*, as one book title proclaimed.

And it was increasingly *men* who were responsible for household alcoholic beverage production in the Chesapeake. By the late eighteenth century, men were sharing the tasks of production by exchanging ideas, ingredients, tools, and drinks. Historians who have studied the masculinization of other forms of labor, such as delivering babies and dairying, have concluded that men took over these arenas because of misogyny or anxiety. However, the shrinking of women's roles in making alcohol was not due to misogyny or male anxiety alone. It was a combination of men's enthusiasm for newly scientific forms of production, the increasing density of the population, and the requirements of supplying the Continental Army with liquor during the American Revolution that led to the re-gendering of alcohol production during the late eighteenth century. In particular, the ban by the Quartermaster's Department on women selling alcohol to the army and the official shift in rations from (women's) ale and cider to (men's) rum in 1781 cemented the notion that making alcohol was a science that belonged within men's domain. Women supported this masculinization of labor by purchasing the liquors that men distilled — happy, in some cases, to forget that making alcoholic beverages used to be their responsibility.[5]

Overconsumption of alcohol was perceived to be a problem in the Chesapeake for the first time during the late eighteenth century. The increasing availability of tea and coffee, for those who could afford them, made drinking alcoholic beverages appear a willful choice. Most studies of rising concern about alcohol focus on the influence of New England moralists, a story that does not suffice for all parts of the country. In the Chesapeake, white planters' fears of slave uprisings caused them to grow worried about slaves' drinking, creating the conditions for the southern temperance movement that would take root later. Slaves and servants in turn contested the power of the planters by making, stealing, and drinking alcoholic beverages. Thus, alcohol ultimately became a problem only when wealthier Chesapeake colonists could finally find something else to drink.

A NOTE ON THE TEXT

Quotations from primary sources have been updated with modern spelling, punctuation, and capitalization so as not to distract the reader.

CHAPTER ONE

"It Was Being Too Abstemious That Brought This Sickness upon Me"

Alcoholic Beverage Consumption in the Early Chesapeake

Consider a day in the life of a small planter in the eighteenth-century Chesapeake. He awoke at daybreak and ate a quick breakfast of corn mush and a couple of mugs of hard cider. After several hot, laborious hours of weeding his tobacco field with his son and his servant, and drinking occasional swigs of cider, he returned to his two-room house. Lunch consisted of some spoon bread, some stew, and another mug of cider. If one of the quarterly meetings of the general court in Williamsburg was in session that day, and if our colonist lived close enough to court to attend, he rode to the session to visit with his friends and conduct some business. He stopped every four or five miles at a tavern to water his horse and refresh himself with a bowl of rum punch. When he arrived at court, he met some of his acquaintances, and they went to a tavern to compare tobacco prices and learn the news. Each took a turn buying toddy for the group. A few hours later our colonist returned home, where his wife gave him some boiled beef and apple brandy.

Our small planter drank what sounds like an extraordinary amount of alcohol, and his wife drank almost as much. Their children had their share of alcoholic beverages too, perhaps in a weaker form called small ale,[1] or, more likely, cider. His servant or slave received alcoholic beverages as well. Cider and rum were considered obligatory for servants. Indeed, indentured servants were guaranteed alcohol; the contracts they signed stipulated that planters provide shelter, clothing, food, and alcoholic drinks in exchange for four to seven years of labor.

On average, a small-planter household in the eighteenth-century Chesapeake contained six people, and they consumed an estimated ninety gallons of cider and twenty-one gallons of distilled liquor per year. A large-planter household with twenty or so slaves might drink 450 gallons of cider and 105 gallons of distilled liquor each year. (Historians call such households "upper sort" because the con-

cept of social and economic "classes" was not invented until the nineteenth century.) For example, the family, employees, and one hundred slaves of the Nomini
Hall plantation in the Northern Neck of Virginia consumed 150 gallons of brandy
and four hogsheads of rum annually, costing the plantation's owner £100 per year
independent of the expense for the plantation's cider, wine, and whiskey. While
today the average American drinks negligible amounts of cider, less than two gallons of distilled beverages, and one and a half gallons of wine per year; annual per
capita consumption of alcoholic beverages in early America was fifteen gallons of
cider and three and a half gallons of distilled spirits.[2]

Colonists drank, frequently in same-sex groups, because they came from a tradition of heavy drinking, because there was nothing nonalcoholic to drink, because alcohol offered one of the few ways to dull the pain of illness, and because
alcohol was one of the few pleasures to be had in the early modern world. The
tradition from which most white Chesapeake colonists came encouraged heavy
drinking. During the Middle Ages in England, finding something to drink had
become increasingly problematic. In 1388 Parliament passed a law against throwing garbage into English rivers, indicating that such a practice was common. By
1425, in Colchester, England, the local leather works and surgeon barbers had
polluted the nearby rivers with blood. Rivers and local water supplies in England
bathed the people and their animals, carried refuse and excrement, and teemed
with disease. Insects infested the water, leading prescriptive literature and cookbook authors to offer remedies for the accidental ingestion of water creatures. For
example, writer Richard Suflet recommended drinking large doses of strong vinegar with fleas to cure the illnesses that resulted from swallowing the horse-leeches
that were common in drinking water.[3]

Not surprisingly, once the middling sort could afford ale and cider, they aban-
doned water. "Water is not wholesome soley by itself for an Englishman," warned
sixteenth-century physician Andrew Boorde. "If any man do use to drink water
with wine, let it be purely strained, and then [boil] it; and after it be cold, let him
put it to his wine." By the fifteenth century, water-drinking had acquired the
stigma of poverty, which astonished visitors. For instance, a Swiss sightseer to England in the 1720s cried, "Would you believe it . . . though water is to be had in
abundance in London, and of fairly good quality, absolutely none is drunk? In
this country . . . beer . . . is what everybody drinks when thirsty?"[4]

Alcohol substitutes such as milk, fruit juice, coffee, and tea were generally unavailable to the English. Well-off families who could afford cows usually reserved
the milk to make butter and cheese. Before refrigeration and pasteurization, there
was no way to prevent the juices pressed out of apples, pears, plums, or other fruits

from fermenting from contact with naturally occurring airborne yeasts. Coffee was not introduced to England until 1601 and only became popular with the urban, wealthy sort in the 1650s. It would remain too expensive for the rest of the country until the mid-eighteenth century. Tea did not spread to England until the 1650s, and it remained an expensive luxury item until the tea duties were lowered in 1745. The result was that when the early modern English drank anything, they typically drank alcohol.

And they drank it abundantly. Statistics for consumption of ale in late-seventeenth-century England indicate an annual consumption of 832 English pints, or 999 U.S. pints, per person. To put this figure in perspective, in 2006 annual consumption was just 209 U.S. pints per person in the United Kingdom and 172 U.S. pints per person in the United States. The early modern English drank at meals and at religious and secular celebrations as well as for medicine and for the purposes of health and beauty. Men drank the most, but women and children drank heartily as well. Children, like their parents, drank ale or beer at every meal.[5]

Social gatherings required alcohol. English men, women, and children drank alcoholic beverages sold by their parish churches to raise money during the popular "church ales." English gentry woman Elizabeth Freke described a church ale in her diary in 1713 as "prayers in the church, and with cakes, wine and a barrell of ale." When the Puritans of Dorchester wanted to establish an almshouse, hospital, and school, they built a municipal brewery to raise the money from sales of ale. To offset the expenses of a wedding, a bride's family would hold a bride-ale, at which the family sold ale to the guests. It was customary to drink liberally at funerals, with the poorer sort drinking ale and the wealthier sort drinking wine. A French traveler in Shrewsbury noted the custom, writing that "there stood upon the coffin a large pot of wine, out [of] which everyone drank to the health of the deceased."[6]

Men and women in England and the colonies drank alcoholic beverages at all times of day. In most early modern households, breakfast was a piece of bread or stew with some ale or cider. The account books for the Percy family of Northumberland reveal that in 1512 the Lord and Lady of the manor shared a quart of ale and a quart of wine each day at breakfast. Their two children, ages eight and ten, also shared a quart of ale at breakfast. Dinner and supper included ale, mead, cider, or wine. The accompanying food was often steeped in alcohol. Women constructed sauces from small ale, salt, and vinegar. They pickled hams in strong ale, sugar, and salt and made pancakes with brandy and sack. Wine and brandy preserved fruit and fish.[7]

When Benjamin Franklin worked in a printing house in London in 1724, the other journeymen mocked his unusual water-drinking, calling him the "water American." Franklin, in turn, commented on the Englishmen's beer consumption. "My companion at the press," Franklin reported, "drank every day a pint before breakfast, a pint at breakfast with his bread and cheese, a pint between breakfast and dinner, a pint at dinner, a pint in the afternoon about six o'clock, and another when he had done his day's work" to maintain his strength. It was the English printer's reported habits, not Franklin's, that were typical.[8]

Cookbook writers placed recipes for alcoholic beverages, the most important or popular part of the book, at the beginning. For example, Sir Kenelme Digbie devoted the initial one-third of his popular English cookbook to alcohol recipes; of the book's 337 recipes, 138 were for alcoholic beverages. Authors usually emphasized alcohol recipes in the titles of their cookbooks as well, at a time when the title was the main marketing device in publishing. Digbie's publisher, for instance, titled his cookbook, *The Closet of the Eminently Learned Sir Kenelme Digbie Kt., Opened 1669: Whereby is Discovered several ways of making Metheglin, Sider, Cherry Wine, &c. Together with Excellent Directions for Cookery.*[9]

English men and women drank alcoholic beverages to promote health. Physician Andrew Boorde expressed a typical attitude when he praised wine in 1542 because it "doth quicken a man's wits, it doth comfort the heart, it doth scour the liver; specially, if it be white wine, it doth rejuice all the powers of man, and doth nourish them; it doth ingender good blood, it doth nourish the brain and all the body." Health care providers sold alcohol to their patients: one surgeon in 1633 "obtained a license with a limitation to sell ale to none but his patients." When Elizabeth Freke's maid, Mary Chapman, "complained of her head and back," Freke gave her "meat and drink, coters, cordialls, apothecaries." The belief in the power of alcohol to maintain health and cure illness led Freke to fill her cupboards with forty-two quarts of alcoholic potions.[10]

Wealthy women often drank alcoholic concoctions every day for their health. Freke copied her sister's recipe to "make lodinum" which required "the best sack [wine]," saffron, and "two ounces of the best opium," infused in a "deep earthen pot." Freke's sister wrote that she took laudanum "once a day for almost a year and half. . . . It must be increased so that now I take five and twenty drops in a day." Poorer women relied on ale and cider. When Alice Thornton's mother was sick in 1659, she could not swallow, so Thornton fed her ale through a syringe. London's St. Bartholomew Hospital allotted ill patients three pints of hospital-brewed beer and unspecified amounts of ale daily. Breastfeeding women often drank heavily hopped beer to help their milk run, and forty-three of the medicinal recipes in the first English gynecological handbook contained alcohol.[11]

Medical practitioners not only used alcohol in medicine but taught their patients that periodic intoxication was beneficial because it purged the body of noxious humors. Ale and beer would even stave off runny noses and watery eyes, as one 1708 almanac reminded:

> Keep warm your Body, and your Head and Feet:
> Use Honey in Ale or Beer, with spices sweet,
> And take warm Food, I wish you as a friend
> Lest Pains of Rheums in Teeth or Head offend.[12]

Early modern men and women believed not only that alcohol maintained health but also that it imparted physical strength. Eliza Haywood, a popular advice writer in both England and the colonies, advised servants to drink ale, beer, and cider to maintain their health, warning that it was *tea* that was the "intoxicating spirit." Haywood cautioned that "merely by drinking of tea, which by too much cooling and weakening the stomach, seems to render it necessary to have something warm. You begin with a little, and think you will never exceed a certain bound, but by degrees increase the proportion, you crave still for more, till by frequent use it becomes too habitual to be refrained." Doctors advised drinking distilled spirits during physical labor and in hot weather in order to warm stomachs left cold by perspiration. In England and America, bathing in alcohol was thought to promote strength, and until at least the mid-nineteenth century, most women bathed babies in spirits rather than water. Ill or consumptive adults sometimes bathed in warm white rum.[13]

Beyond using alcohol to maintain health, English men and women often targeted specific diseases. They drank alcoholic beverages mixed with marigolds to cure depression or infused with cowslips to cure memory loss and headaches. Strong beer and eyebright would cure a "weak brain," while beer with catsvalerian roots would cure hysteria. Wives gave their husbands white wine with hops to invigorate them. Well into the nineteenth century, intoxicants were thought to impart physical stamina; drinking deeply asserted virility. Ale and beer would cure melancholy as well. One popular eighteenth-century song advised, "When dull care does attack you / Drinking will those clouds repeal / Four good bottles will make you Happy, / seldom do they fail." In the age before aspirin, alcoholic beverages at least helped to dull the pain of illness.[14]

Men and women believed not only that alcohol would support good health but that it would also make one beautiful. Women applied wines stewed with thyme, sage, winter-savory, sweet marjoram, and rosemary to their faces to maintain fairness, and rubbed their bodies with mead mixed with rose juice and petals.

Authors suggested patting the skin with white wine and flower blossoms to reduce freckles and spots. Women bleached their hair blonde by shampooing it with white wine cooked with rhubarb or dyed their hair red with wine tinted from radishes.[15]

Alcoholic drinks also served as cleaning agents. Prescriptive literature recommended washing plated metal items in wine and then boiling them in spent mash, aquavit, and vinegar (typically made from the second or third pressing of apple cider). Women scrubbed silver with gin, and servants used gin to clean fireplace grates, scouring the inner hearths with sour beer. Molasses, sugar, and sour beer made fresh blacking for fireplace grates. Some writers advised women to rub metal with rum; mirrors with gin. Women cleaned hats and boots with beer or distilled wine to give them a "brilliant jet lustre." Women dipped candlewicks in wine reductions to make them easier to light. There was little that alcohol could not accomplish inside the household. When household items of china or glass broke, brandy or gin mixed with slugs or isinglass (from sturgeon) glued them back together.[16]

Given all of the experience the English had with alcohol, it is not surprising that they created numerous varieties of drink. Most English men and women in the seventeenth century drank ale, produced from wheat and barley, every day. In the West of England where apples grew plentifully, cider was another staple drink. The English brewed perry or mobby from pears, and mead and methelin from fermented honey. Aquavit was a distilled ale, like a whiskey, based on fermented grain. Mum was brewed from wheat; juniper ale was flavored with juniper berries, bay leaves, coriander, and caraway seeds. Buttered ale was ale flavored with cinnamon, sugar, and butter. Cock ale was a mixture of ale and wine, steeped with raisins, cloves, and its namesake, a cooked rooster. The wealthier sort drank wines infused with birch, cowslip, wormwood, cherries, elderberries, gooseberries, quinces, raspberries, damsons, raisins, and roses. Women made syllabub from wine or cider mixed with sugar, nutmeg, and cream. Red hippocras was made of claret, brandy, sugar, spices, almonds, and new milk. Persico was a cordial flavored with the crushed kernels of peaches, apricots, or nectarines. Rum or arrack, an alcohol distilled from the fermented sap of palm trees, was mixed with sugar, citrus juice, water, and spices to make punch. Toddy was a warm mixture of rum, water, and sweetener; flip, another hot drink, was made of two-thirds strong beer, a little rum, some sugar, molasses, or honey, and some eggs or cream. It was stirred with a red-hot rod to make it bubble and to give it a burnt taste. Wealthy Englishmen also drank wines, brandies, and other distilled spirits imported from abroad.[17]

English colonists in the Chesapeake depended on alcohol at least as much as did men and women in England. Alcoholic beverages provided much-needed calories, were thought to prevent scurvy (a property attributed to beer in particular), offered a respite from water that was infected by disease and mosquito larvae, supplied some relief from illness, dulled fear, and helped to foster a sense of community.

Colonists had little choice about whether to drink alcohol. Non-fermented drinks did not yet exist, and they had no way to prevent the juice of apples, pears, plums, and other fruits from fermenting. Tea remained an expensive luxury item until the second half of the eighteenth century, and coffee was unavailable to most colonists until the late eighteenth century.

The absence of markets meant that milk in particular was frequently unavailable. As one colonist explained in 1697, a Virginian had "no opportunity of a market where he can buy meat, milk, corn, and all other things, [he] must either make corn, keep cows, and raise stocks himself" or traverse the countryside to find these items. Only the upper sort could afford cows. Not all cows produced milk, and those that did gave less milk than do modern cows; a cow was considered a "good milker" if she produced over one quart per day. Moreover, colonists did not fence their cows, preferring to devote their labor to growing tobacco. Milk was almost as expensive as *imported* ale and cider in some cases. "Ale [is] 15 [pence] per [large bulbous] bottle and English cyder the same ... milk [is] 3 [pence] per quart," recorded one traveler. Even colonists with access to milk often avoided it because of fears of "milk sickness" caused by consuming the milk of cows that had grazed on wild jimson weed.[18]

As it was in England, the water in the Tidewater was unhealthy at best. The shallow wells that the colonists dug were contaminated easily and bred typhoid fever. Rivers were often filled with refuse, excrement, and insects. As early as 1625, George Percy described the water in Virginia as "at a flood verily salt, at a low tide full of slime and filth, which was the destruction of many of our men." The warm climate of the Chesapeake allowed much of the water to become contaminated with pathogenic bacteria. Malaria pandemics lasted from 1657 to 1659 and 1677 to 1695. During the summer months the Jamestown River stagnated and gave the colonists salt poisoning.[19]

Some areas of the Chesapeake did enjoy relatively safe drinking water. Southern Marylanders in general had clean water in their early years. For instance, St. Mary's City benefited from fresh water. The hilliness of Baltimore and Richmond permitted storm water and domestic waste to drain relatively well. However, even in areas that had safer water, colonists regarded water with suspicion. William Wood was astounded that New England colonists who drank water could "be as healthful, fresh, and lusty as they that drink beer."[20]

Most Chesapeake colonists did not have access to safe water, and areas that began with better water quickly destroyed it. Colonists dumped dead animals, blood, and excrement into the rivers and ponds. In Norfolk the water was brackish and the public spring filled with domestic pollution. In Williamsburg, Ebenezer Hazard warned in 1777 that the water "is very bad . . . which, with the heat of the weather, is sufficient to keep a man in a continual fever." Virginia planter John Randolph reminded his son not to drink water when traveling. "I see by the papers," he wrote, "eight deaths in one week from cold water."[21]

Illnesses and death were often blamed, perhaps correctly, on water drinking. For example, when William Coone sold a slave who died soon after purchase, the new owner sought to be released from the obligation to pay for the now-dead man. Coone refused to excuse the payment, arguing that the new master had allowed the slave to "drink much cold water." When Nicholas Cresswell, a plantation children's tutor, became ill, his doctor advised him "to drink a little more rum than I did before I was sick. In short, I believe it was being too abstemious that brought this sickness upon me at first, by drinking water." A late-eighteenth-century traveler concluded that "during hot weather thirst is so widespread and irresistible in all American cities that several persons die each year from drinking cold pump water when hot." "Printed handbills are distributed each summer to warn people of these dangers," he continued, and "strangers especially are warned either to drink grog or to add a little wine or some other spirituous liquor to their water. People are urged to throw cold water on the faces of those suffering from water drinking, and bleeding is also suggested. Sometimes notices are placed on the pumps with the words: 'Death to him who drinks quickly.'" Cautions against drinking water were published widely, and in 1766 *The Virginia Almanack* humorously warned:

> Now we advise that you would not,
> Drink water when you're very hot;
> For doing so, the blood congeals,
> And throws the grease into your heels.[22]

Fears of the effects of water drinking lasted throughout the eighteenth century. When Philip Mazzei, an Italian immigrant, asked for a glass of water at a dinner party when visiting Thomas Jefferson, "the host, next whom I sat, whispered in my ear, asking with a smile if I could not drink something else, because the un-expected request for a glass of water had upset the entire household and they did not know what they were about."[23]

"There is unquestionably too much spirituous liquors drank in the newly set-tled parts of America, but a very good reason can be assigned for it," traveler John Melish declared in the early nineteenth century. "The labor of clearing the land is rugged and severe, and the summer heats are sometimes so great that it would be dangerous to drink cold water." His concerns were shared by others. A tourist in the South in 1822 recorded in his journal that "we now most seriously felt the effects of drinking the water . . . which . . . acts as a cathartic." He noted as well that many found water unpleasant to drink, writing that "to a person unaccus-tomed to the water, its taste is insipid in the extreme."[24]

Chesapeake colonists who regarded water with suspicion drank alcohol with hope, sharing English beliefs in its benefits. Planter Landon Carter treated both his daughter and his slaves with alcoholic concoctions. When his daughter, Judy, was sick in 1757, Carter treated her with a "weak julep of rum with salt tartar and pulvis castor" and recorded that after Judy drank the mixture "a small fever en-sued which wore away by night gradually and the child mended." At other times when Judy was ill, Carter prescribed her "toast and beer." When Carter's female slave, Charlotte, fell ill, Carter ordered that "some bread and beer to be boiled and sweetened for her."[25]

Other colonists applied spirits directly to the skin. For example, Hannah Huth-waite recorded a recipe "for a Syatick Pain" in which the sufferer was instructed to apply a mixture of aquavit and vinegar to the painful site. Lucious Bierce ap-plied alcohol to his body when he felt unwell in Virginia, rubbing his sore feet with whiskey to restore them. James Gordon "anointed" his strained ankle with brandy, "which eased it." Rum, wrote traveler Edward Ward, was "adored by the American English . . . 'tis held as the comforter of their souls, the preserver of their bodies, the remover of their cares, and promoter of their mirth; and is a sov-ereign remedy against the grumbling of guts, a kibe-heel [chilblain] or a wounded conscience, which are three epidemical distempers that afflict the country."[26]

Many Americans continued to use alcohol in pursuit of health well into the nineteenth century. Prescriptivist Lydia Maria Child rebuked the "many [who] think distilled liquors are necessary for those who work in cold, damp places" in 1837. Even when colonists went to Virginia's mineral springs to take the "water cure," they continued to drink alcohol. For instance, when Robert Mylne went to the springs in 1774, he drank wine and alcoholic punch throughout his visit.[27]

Some colonists believed that alcohol could cure ailing animals as well. When

Doctor Smith's Boy's horse was "taken with the lock jaw," her owner "poured liquor and laudanum into her," although to no avail. Landon Carter had better luck when he gave his cow "with the blind staggers" three doses of warm beer with rattlesnake root, after which the cow "got pretty well and feeds about as usual."[28]

Much of colonists' consumption was, in a sense, gendered. White men often drank in groups out-of-doors at public festivities, church services, and court days as well as at the militia musters and elections, activities limited to men only. White women more typically drank indoors at sewing circles, birthings, and meals. Men had many more opportunities to drink because they were able to leave their homes far more easily: even free women needed to stay in their houses to nurse and raise children as well as to prepare daily meals.

Male colonists drank at outdoor celebrations such as royal holidays. Officials of Rappahannock County, Virginia, spent ten thousand pounds of tobacco on alcoholic beverages and public festivities in 1688 to celebrate the birth of King James II's son, James Francis Edward. (Two years later the county built a jail for only six thousand pounds of tobacco.) When King William III died, the royal governor of Virginia celebrated the ascendance of the new Queen with cannon salutes and glasses of rum and brandy for the men who attended the celebration in Williamsburg. Similar celebrations caused William Byrd to remark that "there was a great noise of people drunk in the street [a] good part of the night." Refraining from drink on a royal holiday was so unusual that Byrd recorded one instance of it in his diary. "The Governor had made a bargain with his servants," he reported, "that if they would forbear to drink upon the Queen's [Anne] birthday, they might be drunk this [the next] day. They observed their contract, and did their business very well and got very drunk today." When surveyors located new land and claimed it for Governor Spotswood in 1716, they "loaded all their arms and we drunk the King's health in champagne, and fired a volley; the Prince's health in burgundy, and fired a volley; and the rest of the royal family in claret, and fired a volley. We drunk the Governor's health and fired another volley. We had several sorts of liquors, namely Virginia red wine and white wine, Irish usquebaugh, brandy, shrub, two sorts of rum, champagne, canary, cherry punch, cider." Toasting the wrong personage, on the other hand, could get a colonist in trouble. When three men were found "drinking the Pretender's health," they were fined the enormous sum of twenty pounds each.[29]

Men routinely drank in church during the eighteenth and nineteenth centuries. In 1710 Virginia's governor signed a law to restrain drinking in churchyards. The habit of drinking in churchyards was too ingrained for the law to be effective, however. When "my brother Custis and I and Mr. Dunn walked into the churchyard," Byrd noted, "we saw several people drunk, notwithstanding the late law." A minister in South Carolina in 1746 noted that colonists left the service to drink rum punch, returning to church only after they had drunk their fill. When William Byrd led a group of men to determine the dividing line between Virginia and North Carolina, he wrote that "we shall be provided with much wine and rum as just enable us, and our men to drink every night to the success of the following day."[30]

Men also drank in court. It was customary for juries and judges to consume alcoholic beverages during court sessions. Taverns frequently were built next to the courthouses, or court was held in a tavern itself, for the convenience of court officials. For example, in 1647 the Lower Norfolk Court awarded Lawrence Phillips a license to operate a tavern next to the courthouse "for the accommodation of the courts." As a result of the proximity of drink, Byrd recorded that when one "walked to the courthouse . . . the people were most of them drunk." One desperate judge finally informed a jury of men (women were not allowed to sit on juries) who had been unable to come to a decision that "they would not be permitted to eat or drink until reaching a verdict." Judges rewarded juries with a round of drinks for each verdict. Virginia's quarterly court meetings were opportunities for men to drink wildly. During court days in Williamsburg, Byrd recorded that "we drank some of Will Robinson's cider till we were very merry and then went to the coffeehouse and pulled poor Colonel Churchill out of bed." One planter "appearing drunk and insolent" in the Williamsburg courthouse told the "undersheriff to kiss his arse in the face of the court." The judge forgave him three weeks later after he begged the court's pardon. Richard Patterson was likewise drunk when he stepped into the same courtroom and shouted, "Come here you dogs and fight!" He was remanded to the sheriff's custody until he became sober.[31]

The legislature required white men to drill with a militia in case of Indian attacks, and the resulting militia days offered another chance to imbibe. Byrd "caused a hogshead of [rum] punch to be made for the people when they should come to muster" for one such occasion. On another, Byrd recorded, "After 3 o'clock we went to Colonel Randolph's house and had a dinner and several of the officers dined with us and my hogshead of punch entertained all the people and made them drunk and fighting all the evening, but without much mischief." Robert Wormeley Carter noted that he had treated "both Capt. Peachey's and Griffin's companies at the church" with twenty gallons of rum. Alcoholic beverages were such an intrinsic part of the militia muster that boys playing "militia" ended their games with rounds of drinks.[32]

Elected militia officers treated their men to alcoholic beverages as part of their positions. One newly elected colonel promised, "I can't make a speech, but what

I lack in brains I will try and make up in rum." Mustering men mixed some of their brandy charcoal, saltpetre, sulfur, cobine nitre, and brandy to make gunpowder. Men who enlisted expected and received daily alcoholic beverage rations, with extra rations on battle days. One woman watching soldiers in 1775 recorded that "constant draughts of grog supported them. . . . I know that they were heated with rum till capable of committing the most shocking outrages." Continental troops at Valley Forge during the impoverished months from December 1777 to February 1778, a time when the army was not in battle, drank over 500,000 gills of rum and whiskey.[33]

Candidates standing for election similarly provided alcoholic beverages to supporters, inevitably male, at polling sites. George Washington spent over £37 on "brandy, rum, cider, strong beer, and wine" for freeholders on election day in 1758 in order to win a seat in the House of Burgesses. When Virginia colonist Robert Munford wrote *The Candidates,* his mid-eighteenth-century play about electioneering, candidates running for office drank and gave away so much alcohol that Munford named his characters "Sir Toddy," "Mr. Julip," "Capt. Punch," and "Guzzle." Munford emphasized the role of alcohol in his prologue: "When strove the candidates to gain their seats / Most heartily, with drinking bouts, and treats." Successful candidates naturally drank while in office. "My son and I visited the Congress," one tourist recorded in 1794, "when Colonel Parker left his bench and went to refresh himself at a table where there were earthen pots and bottles of molasses liquor for the members."[34]

Other male-focused public events such as horse races and auctions depended on alcoholic beverages. One attendee wrote that at a horse race "a number of people collected there to see quarter races . . . grog as usual had great effect upon them and created much noise." Shopkeepers gave drinks to entice potential customers, and the auctioneers passed alcoholic drinks to those who bid. Williamsburg merchant Joseph Scriviner regularly provided his customers with complimentary alcoholic beverages. When Scriviner died in 1772, his obituary commented that his death would be "often regretted by the lovers of a social evening and a cheerful glass."[35]

Chesapeake women's general illiteracy (girls attended schools or tutors less frequently and for less time than did their brothers) means that women left meager records of their drinking customs. Still it is clear that women had a few gatherings of their own that included imbibing. When women collected to sew or quilt, there was alcohol. "Mrs. Garlick and Sally came," adolescent Frances Baylor Hill noted in her diary, and "we spent the day agreeable eating drinking and quilting." Soon-to-be mothers drank during childbirth to help mitigate the pain of labor.

Their female neighbors gathered before, during, and after the birth, called the "lying in" period. The pregnant woman was expected to have prepared for the women's visits by having alcohol and cakes on hand. "I sent Mr. Harvey a bottle of wine," noted Byrd, "for his wife who was ready to lie in." Some women used alcohol in childbirth in other ways. One African-American midwife had her patient "put a spoonful of whiskey in her left shoe every morning to keep evil spirits from harming the unborn child." Popular songs in the eighteenth century advised women that "'tis safer for ladies to drink than to love," and given the dangers of childbirth, such songs might have been correct.[36]

Several occasions permitted men and women to drink alcohol together. The most important might have been at mealtimes. Colonists often had a drink before going to sleep: "a good hearty cup to precede a bed of down" was how one traveler described it. And at least the wealthy woke to a breakfast with "the cold remains of the former day, hashed or fricasseed; coffee, tea, chocolate, venison-pastry, punch, and beer, or cider, upon one board," as a family guest described it. "I gave some bread and butter and cheese and some strong beer for breakfast," wrote Byrd. Indentured servant John Harrower wrote that his "method of living" included alcoholic beverages at every meal. For breakfast and dinner he had ale, bread, and cheese; with lunch he added meat. "At 12 o'clock," the Virginia plantation tutor explained to his wife in 1774, "I have as much good rum toddie as I choose to drink, and for dinner we have plenty of roast and boiled [meat] and good strong beer." On cold mornings, Harrower and the family "drank a small dram of rum made thick with brown sugar for the cold." A late-eighteenth-century visitor to Virginia explained that at a plantation house "as soon as you get up, your host may possibly invite you to partake of his julep, which is a tumbler of rum and water, well sweetened, with a slip of mint in it." Finally, men and women both drank at the popular outdoor meal called a barbeque, "an entertainment" that, as one traveler described, "generally ends in intoxication."[37]

Funerals and weddings also saw both men and women drinking alcohol. It was the responsibility of the deceased to have left funds to provide the guests with alcohol, or the cost of the drinks would be subtracted from the estate. "We walked to Mrs. Harrison's to the funeral," recorded Byrd, "where we found abundance of company of all sorts. Wine and cake were served very plentifully." Drinking at funerals could be prodigious, and the executors of planter John Grove's estate subtracted at least one thousand pounds of tobacco for the cost of the liquor at his funeral. Drinking at weddings could be even more substantial. When Durand de Dauphine visited Virginia in the 1670s, he complained that the guests had become so drunk at a wedding that he had difficulty locating a place to sleep free from drunken, stumbling bodies. Even in the morning, he could "not see one who could stand straight."[38]

Although it is not yet possible to know whether slaves' alcoholic beverage consumption was similarly gendered, they certainly consumed alcohol, as had their predecessors, indentured servants. As planters moved from servants to slaves after 1680, they continued some of the traditions of providing alcoholic beverages, and slaves worked to enforce those traditions. Just as white men and women drank at holidays, so too did slaves. Slaves and servants drank at the three- to five-day Christmas holiday that most planters permitted, and slaves expected their masters to provide alcohol for the celebrations. One traveler described a slave holiday in Queenstown, Maryland: "100s and 100s of blacks were assembled—wonderfully interspersed with whites young and old—gaming, fiddling, dancing, drinking, cursing and swearing formed one of the most tumultuous scenes I ever beheld." Planters gave their servants gifts of alcohol at this time as well. John Harrower recorded that as an indentured servant at the Belvidera plantation he received "two bottles best rum and some sugar" on Christmas day in 1774. "Mammy Harriet," a slave on Virginia's Elmington plantation, is said to have recalled receiving whiskey at Christmas, and, after the American Revolution, on the Fourth of July as well. "He did not miss givin' us whiskey to drink," the owner's wife later recorded her saying of the plantation owner, "a plenty of it, too."[39]

Planters also supplied slaves and other workers with alcohol for particularly onerous labor. At harvest time it was customary to give slaves alcohol that planters often purchased expressly for this purpose. Thomas Jefferson bought thirty gallons of whiskey annually for his slaves to consume during harvest, and Virginia planters Richard Eppes and William Jerdone similarly gave their slaves whiskey during harvest. Slaves on one of South Carolinian Henry Laurens' plantations drank over thirty gallons of rum during three months of harvest in 1766. One scholar has concluded that during harvest most laborers received a half-pint to a pint of rum per day. When Virginian John Blackford ran out of whiskey at harvest, his slaves "moved a little stiff as they had no bitters the whiskey has given out which they all love dearly." Mammy Harriet allegedly recalled that at harvest there were "two gre't hamper-baskets full o' bottles o' whiskey,—a pint for ebery man an' half a pint for ebery 'oman . . . de young gals alays tik de whiskey . . . an' put water an' sugar in it." Planters who hired white servants in the seventeenth century and early eighteenth century found that the alcohol they consumed accounted for 37 percent of their employment expense.[40]

Planters gave slaves alcoholic beverages as a reward for extra work and at unexpected times to curry loyalty. When an "abundance of snow fell in the night,"

Byrd recorded, "my people made paths for which I gave them cider." "I gave all the people a dram," he noted on another occasion, "after planting in the rain." "At night talked with my people," Byrd wrote, and "gave them cider, and prayed." Another day, Byrd reported that he had "inquired of my people how everything was and they told me well. Then I gave them some rum and cider to be merry with." The daughter of an early-nineteenth-century Virginia planter recalled that "fodder-pulling was looked on with dread by most planters, as the hot work among the corn-stalks gave the negroes chills and fevers." She remembered that her father "guarded his negroes against sickness by providing two barrels of whiskey for this season. Every man and woman came for a cup of it when the day's work was over."[41]

Slaves used alcohol at births, weddings, illnesses, and funerals, most of it supplied by planters. Peter Jefferson, Thomas Jefferson's father, allotted his pregnant slaves a quart of brandy to drink during childbirth. Mammy Harriet, of Virginia, recalled that "I had a weddin' . . . marster brought out a milk-pail o' toddy an' more in bottles." Joseph Ball instructed the nephew who was managing his plantation to "let the cider be beaten in as good order as can be, without the tobacco's suffering; and let the negroes have what will do them good; but none for feasts; unless it should be for a burying or wedding; and then but sparingly. . . . You must give the negroes at each plantation a bottle pretty often in the winter; and when they are sick." Mammy Harriet recalled that "a disease called black tongue appeared among the negroes," and "not knowing what medicine would check the disease, he [the master] resolved to give none . . . a liberal use of port wine . . . justified his hopes and expectations. He did not lose a case." Slaves expected alcohol at funerals as well. When a slave came to planter Robert Carter "to buy brandy to bury his granddaughter," Carter gave the slave a dollar, "telling him he might lay out his annuity as he pleased." And Henry Knight, who visited the Chesapeake in the early nineteenth century, noted that slaves often drank the dead to their new homes.[42]

There is considerable evidence that planters wanted to cut back on the alcohol they gave to slaves. George Washington considered halting the practice of providing spirits during harvest but concluded that since his slaves "have always been accustomed to it, a hogshead of rum must be purchased." Landon Carter provided his slaves with only one shirt, instead of the customary two, to force "them to buy linen to make their other shirt instead of buying liquor with their fowls." Some planters did succeed in eliminating or reducing alcohol rations by the late eighteenth century. At this point, as Polish visitor Julian Ursyn Niemcewicz phrased it, Virginia planters gave their slaves, "only bread, water and blows."[43]

Still, according to accounts of the time, slaves managed to obtain enough al-

cohol to get drunk recurrently throughout the seventeenth, eighteenth, and nineteenth centuries. Planters recorded slave drunkenness frequently. "I reproved
George for being drunk yesterday," noted Byrd. "I was out of humor with my wife,"
wrote Byrd on another occasion, "for trusting Anaka with rum to steal when she
was so given to drinking, but it was soon over." "My man John got drunk," Byrd
jotted another day, "for which I reprimanded him severely." Slaves were able to
steal alcohol because many planters allowed them either explicitly or implicitly
to roam at night. James Tayloe had to ask Landon Carter to crack down on slaves'
night wanderings, complaining that "my people are rambling about every night"
at Carter's plantation.[44]

Sometimes slaves stole drink, and masters tolerated it to a degree. George
Washington considered promoting his slave Cyrus to gardener if he could give up
some of his drinking, "to be sober, attentive to his duty, honest, obliging and
cleanly," a suggestion that planters could forgive slaves for stealing alcohol. Washington later concluded angrily that his slaves stole two glasses of wine to every one
that he had the chance to consume. Moore Fauntleroy told Landon Carter that
his house had been robbed of several gallons of rum, that he suspected his enslaved groom, and that he did not propose to take any action about the incident.[45]

Slaves who stole alcohol too frequently incurred the wrath of their owners. In
1721, "Frank, a negro man belonging to Henry Long" was presented to the Richmond Court for "feloniously and burglariously [sic] break[ing] and enter[ing] and
[stealing] ten gallons of rum of the value of forty shilling sterling." James Mercer
sold his house servant, Christmas, because of her fondness for liquor. William
Byrd "discovered that by the contrivance of Nurse[,] Anaka Prue got in at the cellar window and stole some strong beer and cider and wine. I turned Nurse away
upon it and punished Anaka. . . . In the afternoon I caused Jack and John to be
whipped for drinking at John [Cross's] all last Sunday."[46]

Slaves also bought alcohol when they could, as well as the ingredients to make
it. Skilled slaves who were rented out were often allowed to keep part of the money
that they earned. With this money, slaves could buy drinks in taverns where the
tavernkeeper permitted it, despite laws prohibiting such transactions. For instance,
neighbors concluded that at least one tavernkeeper in Elizabeth City County "en-
tertains and harbours negroes and idle, disorderly persons." Some overseers sold
alcoholic beverages to slaves, as did some farmers. Landon Carter's waiting man
and slave doctor, Nassau, obtained alcoholic drinks from a nearby farmer. Slaves
on Coffin's Point plantation were angry when they discovered that the molasses
they had purchased from another plantation superintendent was watered down.
The Scottish factor stores, which sold goods to small planters in exchange for their

tobacco and which spread widely after 1760, gave slaves the opportunity to purchase more alcohol. A detailed study of one such store reveals that 10 percent of
that store's customers were slaves and that their most frequent purchase was alcohol. In the late eighteenth century, Chesapeake colonists developed marketplaces;
slaves would sell there and make purchases with the proceeds. One traveler to
America concluded that slaves "generally return drunk from market."[47]

It is clear that colonists, white and black, male and female, drank copiously.
Indeed, alcoholic beverages accounted for an average of at least 10 percent of di-
etary expenses. Hard liquor was on average 45 percent alcohol. Beer was likely
around 5 percent alcohol, hard cider around 10 percent alcohol, and wine around
18 percent alcohol. Beer consumption, however, was negligible in the colonial
Chesapeake. Colonists drank mostly cider and a distilled cider called apple brandy.
Even after whiskey increased in popularity after 1800, per capita consumption of
hard cider remained fifteen or more gallons per year. Wine consumption was
light. The typical American drank only one-tenth of a gallon of wine annually in
1770. One historian has concluded that in 1770 annual per capita intake of alcohol (not alcoholic beverages, just the pure alcohol) from all sources was 3.5 gallons.[48]

Of course, not all colonists drank equally. White adult men probably drank
two-thirds of the total distilled sprits consumed. Women, children, and slaves
were inhibited by social conventions, physical stature, and planters' concerns.
Wills dating from 1740 to 1790 indicate that white widows usually kept three or
four barrels, or about fifty-four gallons, of cider per year for themselves.

It is harder to determine how much alcohol slaves consumed. Remarks on
slave alcohol consumption are vague; many are so tainted in racism that they are
difficult to interpret. For instance, "A Planter" published an anonymous article
stating that "every planter knows there are many negroes . . . [that] are always
found ready to barter away their whole weekly allowance to some neighboring
dram shop for a gallon of whiskey." Other accounts are problematically imprecise.
One early-nineteenth-century newspaper article on slave management suggested
only that "in the heat of the day I give them [slaves] some whiskey to drink . . . and
give them as much meat as they could eat during the warm weather."[49]

Still, it would be fair to conclude that a typical small-planter Chesapeake
household with six people — one surviving parent, two children, and three tithables (servants or slaves for whom masters paid a property tax) — required 75 gallons of cider and about 17.5 gallons of distilled liquors per year. A representative
large plantation with thirty people — say, twenty-two slaves, an overseer, two indentured servants, a couple of day laborers, two adults, and a child — needed 450

gallons of cider and 105 gallons of distilled spirits per year. Since slaves received less alcohol than did whites, even if their allotment of cider was only one pint per day, the plantation still consumed 72 gallons of cider per year and 25 gallons of liquor.

But Chesapeake colonists were living in a wilderness, three thousand miles from their mother country of England. How were they going to get the alcohol that they required?

CHAPTER TWO

"They Will Be Adjudged by Their Drinke, What Kind of Housewives They Are"

Gender, Technology, and Household Cidering in England and the Chesapeake, 1690 to 1760

For centuries in England the task of producing alcoholic beverages had belonged to women. Indeed, in 1656 Englishman John Hammond denounced Chesapeake women for making insufficient amounts of alcohol, particularly unhopped corn brews that he called "beer," for their households. Hammond reported that in Virginia and Maryland "beer is indeed in some places constantly drunken, in other some, nothing but water or milk and water or beverage; and that is where the goodwives (if I may so call them) are negligent and idle." These "slothful and careless" women, Hammond announced, had reduced some of the local men to drinking water. Hammond instructed women to increase their production. "I hope this item will shame them out of those humors," Hammond concluded, and he reminded women that "they will be adjudged by their drinke, what kind of housewives they are." As in England before 1700, making alcohol was women's work.[1]

When the Richmond County, Virginia, court sentenced Thomas Phillips to ten lashes, "well laid on," at the public whipping post for "picking open the lock of George Sisson's chamber and taking out the keg of his cider house and disposing of his cider" in 1722, the judge thought nothing of describing the cider as George Sisson's property, even though it was very likely made by his wife. The Richmond Court's ruling shares a close connection with Hammond's critique of "slothful" housewives. By the late seventeenth century, when women in the Chesapeake were fulfilling Hammond's challenge and regularly making the alcoholic beverages necessary to sustain their households, people in England were concluding that alcohol production was *men's* work. Chesapeake men and women came to depend on female-produced ciders and unique New World drinks rather than on the new male-produced beer that was increasingly popular in England. So, when the Richmond Court described the cider as George Sisson's property, the court unintentionally disguised the drink's true origins: the women of Sisson's household.[2]

Women's cidering activities, hitherto unnoticed, reveal the delayed and rural nature of the early Chesapeake in the context of the English colonies and the Atlantic world. Alcohol production in the Chesapeake increasingly resembled that of rural England in the sixteenth century, where women produced cider and unhopped ale. The study of alcohol illustrates how out of step the early Chesapeake was in the Atlantic world, albeit for very practical reasons.

Small-planter households rejected imported scientific and technological innovations. While Western Europeans made alcoholic beverages that required early modern technological innovations, most Chesapeake households continued to produce only raw, simple, naturally fermenting ciders. Western Europeans making beer used the hops flower, a new discovery, and they distilled liquors with newly invented stills, whereas most Chesapeake colonists made only ciders from apples, peaches, and persimmons; concoctions of molasses and water; a little unhopped ale from corn and molasses rather than from oats and barley; and some apple brandy distilled from cider in primitive stills. It is worth noting that southerners would not generally enjoy the beverages now called beer, ale, or lager until the spread of refrigeration in the twentieth century. They were also unsuccessful at producing wine and frustrated at importing drinks. Instead of increasing their participation in Atlantic world trade, Chesapeake colonists chose to make their own cider.

This Chesapeake isolation allowed colonists to decline to engage in the masculinization of alcohol production that prevailed in other Atlantic world countries. Whether in Latin America, Africa, or Western Europe, the task of making alcohol became men's domain in the sixteenth and seventeenth centuries. For example, in colonial Mexico, women had long brewed *pulque*, a cactus wine made from maguey plants. In the Andean region, indigenous women brewed a corn liquor called *chicha*. But in the early seventeenth century, Spanish newcomers desiring to control the indigenous population declared that alcoholic beverages were the domain of the Spaniards. Despite local women's resistance, Spanish men assumed control of production, distribution, and consumption of alcohol in much, if not all, of Spanish America in the seventeenth century.

Similarly in Africa, men and women on the Gold Coast (present-day Ghana) had cooperated in distilling *akpeteshie*, a palm wine, for centuries. It was an *Akan* taboo for a woman to work with the palm trees, but it was unacceptable for a man to distill alone. In these regions, an African marriage was finalized only when a bride and groom's kin gave the bride the right to make beer for her own and her husband's ancestors. However, Dutch colonization and increasing urbanization and commercialization reduced women's role in distilling in coastal Africa until making alcohol became an activity performed solely by men by the 1800s. Holland, Germany, and England all saw men assume control of alcoholic beverage production between the fourteenth century (Holland and Germany) and the seventeenth century (England).

Chesapeake colonists swam against this tide of masculinization and expansion in scale. While men in Europe, and even in the fellow colonies of New England and the Middle Colonies, claimed alcohol production as their domain, women in the Chesapeake actually expanded their cidering. This shift was the result of three factors: the Chesapeake's immigration patterns, its widely scattered population, and its tobacco monoculture.

Men who owned one to two hundred acres in the Chesapeake were considered small planters. Their households included an average of six members, two or three of whom were indentured servants or, increasingly, slaves. Small planters raised crops for a living and often worked their land themselves. Their households made up between two-thirds and three-quarters of Chesapeake households from the late seventeenth century on.

Producing alcohol for small-planter households was women's province from the seventeenth until the latter half of the eighteenth century. In these households, women made enough raw alcoholic beverages — undistilled ciders of apples, peaches, and persimmons — to meet their consumption needs in autumn and winter. Women also sometimes made molasses "beer," a fermented mixture of molasses, water, and yeast. When small planters' supplies of raw alcoholic beverages ran out, they depended on large planters to sell them their surplus distilled liquors, the subject of the next chapter. Smaller households could not afford the wealthy planters' expensive technology, particularly the distilling equipment that provided the longer-lasting alcoholic beverages that the men of England and New England claimed as their domain. Small-planter households purchased very little imported alcohol, so they depended on women's cidering.

Since Chesapeake colonists built few towns, marketplaces, or shops, colonists had to produce or import whatever goods they needed. Importing alcohol was frustrating. Bottles, kegs, and barrels consistently broke or leaked during transportation. "Your convict ship arrived safe with the goods, if one may call that safe where everything is damaged and broke to pieces," fumed William Byrd. "Opened Mr. Lee's claret," wrote Landon Carter, "and had 5 bottles broke in it out of the 6

An examination of the frontispiece of an advice book for women on "preserving, physick, beautifying and cookery" reveals in the top panel that making alcoholic beverages was women's work even though the usual historical sources do not mention their efforts. Hannah Woolley, *The Accomplisht Ladys Delight* (London, 1684), reproduced by permission of the Folger Shakespeare Library.

dozen, entirely by the loose careless way of packing it up." "The corks were so thick and rotted," John Custis recorded, that "they all came off," and the liquid spilled. "We unpacked the beer that came from England," noted Byrd on another occasion, "and a great deal was run out." And again: "Much of my wine was run out," sighed a resigned Byrd, "God's will be done." Small-planter families elected not to import alcoholic beverages themselves, given its costs. Instead they made their own.[3]

Traditionally, the task of making alcohol fell to women as part of cookery. English women had been responsible for producing alcoholic drinks since at least the third century. While English women's brewing tasks differed by class, with the Countess of Shrewsbury directing her female servants in brewing, while a shepherd's wife would brew her own, they were responsible for cidering, brewing, flavoring wines with berries and fruits, making mead, and distilling liquor from garden fruits and berries. Sixteenth- and seventeenth-century writers of prescriptive literature agreed that producing alcohol was women's work. Gervase Markham was writing of the malthouse when he reminded men that "this office of place of knowledge belongs particularly to the housewife; and though we have many excellent men-malsters, yet it is properly the work and care of the woman, for it is a house-work, and done altogether within doors, where generally lies her charge." "The art of making the malt," Markham continued, "is only the work of the housewife, and the maid-servants to her appertaining." Anthony Fitzherbert stated in his "prologue for the wives occupation" in his popular *Book of Husbandry* that "what works a wife should do in general" were "to order corn and malt to the mill, to bake and brew." Thomas Tusser's widely circulated *The Good Huswife* told housewives that at dawn they should "set some [female servants] to grind malt" and that "where brewer is needed be brewer thyself, what filleth the roof will help furnish the shelf." Tusser reminded women that, "well brewed, worth cost: one bushel well brewed, out lasts some twaine, two troubles to one thing is cost no gain." He warned that in making ale, "too new is no profit, too stale is as bad: drink sour or dead makes husband half mad." The word "brew" is a distant etymological relation of words such as "broth" and "boil," a sign that brewing was part of cookery and thus women's work.[4]

English cookbooks both reflected and promoted the experience of women as alcohol producers by providing instruction *for women* in methods of brewing, distilling, and cidering. Richard Bradley, author of the popular *The Country Housewife and Lady's Director*, subtitled his book *Containing Instructions for Managing the Brew House, and Malt Liquors in the Cellar; the Making of Wines of all Sorts.* Bradley's books included lengthy discussions on brewing that recommended the

Frontispiece of an advice manual on making jellies, candying fruit and flowers, and baking cakes, showing that elite women in England traditionally distilled alcoholic concoctions called "waters" that relied on nuts, berries, and fruits. Hannah Woolley, *The Ladies Directory in Choice Experiments and Curiosities of Preserving* (London, 1662), reproduced by permission of Special Collections, University of Glasgow.

best water, corks, vessels, and cellars. Bradley urged women to keep bees so they could make mead from the honey. He also exhorted them to cultivate gooseberries, elderberries, and currants for home winemaking. He explained that "the fair sex" should "have care and management of every business within doors, and to see after the good ordering of whatever is belonging to the house."[5]

Gervase Markham's *Country Contentments and The English Huswife* likewise directed its advice to women. Mary Cole, author of *The Lady's Complete Guide*, reminded women that "the house-keeper cannot be said to be complete in her business, without a competent knowledge in the art of brewing." Sir Kenelme Digbie, in his best-selling cookbook, advised women on how to cider, suggesting that they "take a peck [eight quarts] of apples, and slice them, and boil them in a barrel of water . . . draw forth the water . . . three or four times a day . . . then press out the liquor, and tun it up; when it has done working, then stop it up close." He never described how to "tun" or "draw" the cider, since he could assume that women already had this knowledge.[6]

Since most free colonists to the Chesapeake came from England, it is worth examining English alcoholic beverage production activities in more detail. By the late sixteenth century in London and other densely populated areas, women who did not have the space or the desire to brew found that they could purchase beer. For example, Lady Margaret Hoby brewed her own ale and cider when she lived in rural Hackness, England, but she bought her beer when she resided in London. The brewers there had formed the Brewer's Guild in the fourteenth century with both male and female members. Increasing urbanization and competition led the Brewer's Guild to expand its apprenticeship and membership requirements, and over time women increasingly found themselves unable to join the guild because women, who were expected to marry, raise children, and maintain households, could not spend three to seven years training with a master as the guild required to become a licensed brewer.

The introduction of hops completed the masculinization of alcohol production in England. Dutch aliens, men who were brewers in the masculinized Dutch brewing industry, began to brew and market their hopped beer in southeast England during the fifteenth century. *Beer*, which was brewed with hops, kept significantly longer than *ale*, which, in early modern lexicon, was unhopped. The hopped beer would eventually dominate the market, although its acceptance was slow.

The English continued to prefer the sweeter unhopped ale until at least the end of the fifteenth century. Many English people viewed beer with xenophobia throughout the seventeenth century. "Beer is a Dutch boorish liquor," declared John Taylor in 1651, "a thing not known in England, till of late days an alien to our nation, till such times hops and heresies came among us." By the end of the seventeenth century, however, most English men and women were drinking hopped beer, and this drink increasingly dominated England's domestic alcoholic beverage trade.[7]

The transition to hopped beer held long-term gender implications. The production of hopped beer required more fuel, more labor, and more capital than unhopped ale. Imported hops were expensive, and since hopped beer needed to mature before sale, the producer had to have the capital to maintain larger stocks. The majority of women lacked the capital to brew beer on the commercial scale. Beer and urban alcohol production had become men's domain in England by the end of the seventeenth century.

Men had also assumed control of cidering in England, a topic that scholars have not yet researched. Doctors began prescribing cider to sailors in the late seventeenth century because of its supposed antiscorbutic properties. Landlords in the West Country of England began to produce larger amounts of cider to sell to the navy and to the new "cider houses" in Oxford and London. The commercial ciderers were assisted by Andrew Yarranton's improvements in bottling techniques and John Worlidge's introduction of a portable apple hand-mill, both in the 1670s. These new technologies and new markets combined to favor men, and authors wrote books to aid men in both brewing and cidering. For example, Thomas Chapman's 1762 *The Cyder-Maker's Instructor* taught men how "to make his cider in the manner foreign wines are made . . . fine his beer and ale in a short time." Chapmen, like other authors, offered cider, beer, yeast, and wine recipes for men, because both brewing and cidering had become male fields by the late seventeenth century.[8]

The Chesapeake is remarkable, because there making alcohol remained women's work. In the early seventeenth century, the colonists were "a colony [of] water drinkers" because they had so few women. A few wealthy men deliberately imported wives who had brewing and cidering skills. Allice Burges, for example, was imported in 1621 to be a wife to a colonist. She carried a letter recommending her as "skillful in any country work. She can brew, bake, and make malt &c." Ann Tanner's letter said that she "could spin and sew . . . brew, and bake, make butter and cheese, and do huswifery." Most men, however, did not have wives and so drank water.[9]

The number of women did not equal the number of men until the 1680s. For every white woman who immigrated from London to Maryland in 1634 and 1635, six white men immigrated. From the 1650s to the 1680s, that number dropped, and the Chesapeake had about one woman for every three men. From 1697 to

1707, the ratio reached one woman for every two and a half men. Until the 1680s, men who managed to marry often used their wives to grow tobacco in the fields because of the shortage of alternative labor. The increasing importation of slaves and the decline in tobacco prices in the 1670s and 1680s released white women from raising tobacco to attend to more typical English household tasks. Thus by the late seventeenth century, the Chesapeake contained a rapidly growing population of white women who had more time to engage in cidering and other housework.[10]

The evidence that women were the proper managers of alcohol production in the eighteenth-century Chesapeake is overwhelming despite the fact that no women's journals, diaries, or letters have survived. For example, when Edward Pitway sold his land to George Corke in 1668, he stipulated that half of the fruit the orchard produced each year was to be given to Pitway's wife, Elizabeth, for her cidering. In 1690, when the Accomack County Court awarded land to John Cole, his wife Mary "was to have the privilege during her life to make use of the old orchard [for] making liquor." Henry Baker left his wife, Sarah, "half my land, orchards, etc. during her natural life," for her cidering. Dorothy Needle paid William Downing in cider for teaching her children in 1703; and in 1713, Richard Wharton left "the poor widow Skelton (alias Broadbent) and her daughter" "apples for their use from the orchards." Elizabeth Middleton was brought before the court for selling her alcoholic beverages out of her house without a license in 1723. In 1734, Elizabeth Harrison of Berkely, Virginia, ordered "1 great cooling tub" for her brewing and cidering. And in 1749, John Grymes bequeathed the use and occupation of his house to his wife, specifically including the use of the orchards.[11]

Early-eighteenth-century wills and inventories indicate that women of a variety of economic standings cidered in the Chesapeake. Well-off Elizabeth Ballard of Charles City County owned three cider casks, three cider hogsheads, three cider barrels, and seven dozen bottles as well as two slaves, a feather bed, a mirror, pewter cutlery, some fine cloth, a cow, a horse, and £5 worth of gold. Middling-sort Sabra Crew had two "old cows," one "old horse," an "old gun" without a lock, one "old rug," and one "old blanket," as well as "old pots," "old pot hooks," and "two old tubs" for cidering. Mary Tye, who was of lower-middling status, also owned the tools necessary to cider, although she had little else; a "parcel of old iron," an "old pot with hook," and two "pots with racks."[12]

Room-by-room inventories of households in Williamsburg, Virginia, reveal that cidering equipment was housed in the same spaces where women performed their other household labors, a point that scholars have not noted until now. For example, Ambrose Fielding's household had "3 old tubs, 2 new tubs, 5 cider casks,

a parcel old bottles, 1 great kettle to contain 40 gallons, a skimmer" and two spinning wheels in the "out kitchen." Nathaniel Brandford's 1690 estate contained
two cider casks and three cider pipes, located in the kitchen. John Thompson's
1700 Surry County household also kept "old casks" and "old tubs" for making alcohol in the kitchen. James Burwell's household stored four tubs for cidering in
the kitchen as well. In 1704, Joseph Ring's household in York County maintained
one "old still" in the same chamber as the buttering equipment, such as a "stone
butter pot," thereby combining women's tasks of distilling and dairying in one
room. Joseph Frith's household stored dairying and brewing equipment together
in the dairy as well. Katherine Gwyn kept her syllabub pots in the dairy with her
milk pans, decanter, and butter pots. Arthur Allen's household placed sixteen cider
casks, four pot racks, two cider cloths, six yards of hair cloth (used for straining
cider) and other cidering and molasses-beer brewing equipment including thirty-
five gallons of molasses, in the still house with the spinning wheel.[13]

Cookbooks that were popular among seventeenth- and eighteenth-century
Chesapeake women presumed or asserted the fact that women produced alcoholic beverages. Colonial history scholars have generally ignored cookbooks as
historical documents, but they are a telling source. Letitia Burwell recalled in the
nineteenth century that "every Virginia housewife knew how to compound all the
various dishes in Mrs. Randolph's cookery book." Mrs. Randolph's recipes included mead, ginger beer, spruce beer, and molasses beer. (These "beers" were
unhopped.) Hannah Huthwaite either concocted or copied recipes for quince
wine and raspberry wine in her *Recipe Book*. Martha Eppes had recipes for making grape wine and "instantaneous beer." Frances Parke Custis compiled and created recipes including how "to make cider" in an early- eighteenth-century book
that she later presented to her daughter-in-law, Martha Dandridge Custis Washington. Washington then gave the book to her granddaughter, Nelly Custis, upon
her marriage. The women's careful passing of the cookbook through the female
side of the family suggests that alcoholic beverage recipes were valued property.[14]

Women in small planter households used a laborious procedure to make perry
(a cider of pears), peach "brandy" (a cider of peaches), and persimmon "beer" (a
cider of persimmons), as well as apple cider. Few households had cider presses because they were extremely expensive. Women, and sometimes men, first beat and
pressed fruit in a wooden trough and collected the juice. Then women transferred the remaining pulp to bags of woven hair and drained the pulp. Next they
pressed the remaining pulp through sieves and collected all the juice into a tub,
where it sat covered with cloths or boards for at least twenty-four hours. Women
then decanted the liquid into another container to leave the pulp behind, a time-

consuming process that was repeated several times. The cider was then tunned into cider casks, where it would continue fermenting.[15]

Women who had the time and utensils also distilled alcoholic and nonalcoholic infusions of herbs, berries, water, and wines. For example, Hugh Plat wrote in his 1594 *Divers Chimicall Conclusions Concerning the Art of Distillation* that his book gave "some other necessary knowledges in the art of distillation, concerning such matters as I am assured that every gentlewoman that delighteth in chimical practices will be willing to learn," including recipes for cinnamon water, rose water, and currant wine. Peter Kalm, a Swede who toured America in 1748 and 1749, recorded how "the ladies make wine from some of the fruits of the land. . . . They principally take white and red currants . . . since the shrubs of this kind are very plentiful in the gardens, and succeed very well. . . . They likewise make a wine of strawberries." Kalm also noted that some women rudimentarily distilled the juice that they had pressed out of peaches:

> The fruit is cut asunder, and the stones are taken out. The pieces of fruit are then put into a vessel, where they are left for three weeks or a month, till they are quite putrid. They are then put into a distilling vessel, and the brandy is made and afterwards distilled over again. This brandy is not good for people who have a more refined taste, but it is only for the common kind of people, such as workmen and the like. Apples yield a brandy, when prepared in the same manner as the peaches.[16]

Colonial writers celebrated women's work in making alcohol. "What a useful acquisition a good wife is to an American farmer," exclaimed St. John De Crèvecoeur, "and how small is his chance of prosperity if he draws a blank in that lottery! What should we do with our fruit, our fowls, our eggs? There is no market for these articles but in the neighborhood of the great towns." Colonists, Crèvecoeur stressed, depended on women's skills in food and drink preparation and preservation. The farmer "may work and gather the choices fruits of his farm, but if female economy fails, he loses the comfort of good victuals . . . if we are blessed with a good wife, we may boast of living better than any people of the same rank on the globe."[17]

The question now is *why* Chesapeake women made alcoholic beverages during the late seventeenth and early eighteenth centuries when the rest of the Atlantic world was relying on men. This question is particularly puzzling because immigrants to the Chesapeake in the seventeenth and eighteenth centuries knew that alcoholic beverage production was increasingly men's domain.

The answer is in part demographic. While people from many parts of Europe and Africa emigrated to the Chesapeake, both freely and unfreely, the Chesapeake remained an overwhelmingly English society in the late seventeenth and early eighteenth centuries. The vast majority of seventeenth-century English immigrants left for the New World from urban areas of England, most commonly from the cities of London and Bristol. But most of them really came from rural areas and had traveled to England's cities only recently, looking for work. When they did not find any, they were forced to seek their fortunes abroad.[18]

Chesapeake colonists' rural origins and urban experiences meant that they were familiar with both women's historic rural alcohol production and men's more recent urban production. In parishes, villages, and small towns throughout England, women continued to make small-scale cider and ale until the eighteenth century. Just when women's role in rural English alcoholic beverage production declined and alcoholic production moved toward masculinization during the end of the seventeenth century, English immigration to the Chesapeake slowed. Immigrants to the New World had become familiar with men making their beer and spirits when they lived in London, Bristol, and other urban areas, but retained a memory of women making alcohol, particularly ale and cider, in their rural home parishes.

Chesapeake colonists drew on this memory and depended on women's cider production because the requirements of the tobacco crop left them little time to develop the crops and technology necessary for other forms of male-directed, large-scale, urban alcohol production. Raising tobacco required an extraordinary amount of labor and space. Planters raising tobacco had to clear the land, burn the underbrush, mix ashes into the earth for fertility, prepare and plant the seedbeds, and cover the beds with leaves and boughs to protect them from frost. On warm days, planters removed the coverings and laid them again at night. Planters had to water the plants twice a day if the weather was dry, which required them to haul water in buckets; to weed and thin the plants; to transplant the seedlings as they grew to hills made with hoes; to replace the plants that had died, crush the hornworms that infested the plants, top and sucker the plants; and then to cut, cure, pack, and transport the leaf for sale. If colonists wanted to eat, they also had to shell corn, haul water, slaughter hogs and cattle, salt and smoke meat, replace damaged fruit trees, and chop thirty cords of firewood per cooking hearth. Women had to card, spin, and weave wool; sew clothes; grind corn; nurse their babies; tend their children and, if they could read themselves, teach them to read; make food; preserve fruits, vegetables, and meats; dip candles; press butter and cheese; clean households; plant, weed, thin, and hoe gardens; wash laundry; and, of course, make cider.[19]

Colonists in the Chesapeake chose not to grow the hops and barley or oats necessary for English-style beer, which were labor intensive and time-consuming, at a time when colonists wanted to use their labor, land, and time for raising tobacco. Early modern hops were far more fragile than today's hops and were time-consuming and expensive to grow as well. In order to grow the hops available at the time, planters had to remove the rocks from the ground; plant the hops in deep, mellow soil; shelter them and space them evenly; groom them frequently to prevent mold, lice, and insects; water them often; and protect them from hogs, cattle, and fowl. The vines of the hops had to be twisted, laid with dung, covered, and placed on tall poles. Then the hops had to be picked, cleaned, and turned constantly while drying in bags. In order to replant the following year, the poles had to be stripped, and the land manured. Moreover, as the Royal Dublin (Scientific) Society warned, "It is not proper for poor farmers, or men of small fortunes, to engage far in this improvement, for it requires a considerable stock . . . the expenses will be great, and the undertaker must expect to lay out his money for two or three years, before he can have any return of profit, and even when his hops come to their bearing state, and he is in hopes of making good the charges he has been at, he may be disappointed by a bad season." Furthermore, "the hop is a very tender plant, and an uncertain commodity to deal in, that it is very apt to suffer by winds, blights, mildews, rains, droughts, and insects, and when it wholly fails, the loss is intolerable, and if there be a general good crop, the price will be so low, that it will hardly answer the charge."[20]

The Virginia Assembly offered repeatedly to give a note for 10,000 pounds of tobacco (such notes circulated like money) to any Virginia colonist who raised "so much silk, flax, hops or any other staple commodity (except tobacco) as is worth two hundred pounds sterling." But hops production in the Chesapeake remained insignificant. Colonists focused on the production of the profitable tobacco crop rather than waste their land and their time on a crop that might not succeed.[21]

Another problem in the way of commercial or male-produced alcohol was the lack of markets. Because of the Chesapeake's emphasis on tobacco, it remained overwhelmingly rural, unlike New England and the Middle Colonies, which developed towns. Even small planters in the Chesapeake needed one hundred to two hundred acres for tobacco. In contrast, much of New England was laid out in towns that required all colonists to live within five miles of a church. New England and the Middle Colonies developed markets where colonists could purchase malt, hops, barley, and oats if they did not grow them themselves. A traveler to Winchester, Frederick, and Berkley, Virginia, noted in the late eighteenth century that malt was "little made" and hops "but few made." As a result, beer was "seldom to be met with." The lack of coastwise shipping between the colonies meant that Chesapeake colonists could not import these items from the northern colonies, and the importation of hops from England was negligible because they were expensive and because, given the requirements of growing tobacco, few colonists had time or interest to experiment with them anyway.[22]

Alcohol production in the Chesapeake was irregular. As the Chesapeake languished, commercial brewing boomed in New England. By 1637, Massachusetts had to restrict the number of brewing licenses because the colony had so many men brewing. By 1800, New York had forty breweries, all operated by men, and Pennsylvania had forty-eight. When Ellen Wayles Randolph Coolidge married in Boston, her Virginia mother, Martha Randolph, wrote to her that "I have not sent you those 'receipes' [sic] which as a town lady would be useless to you. For example you need not know how to make soap or blackening or yeast or ley [cheese] and rennet [cheese] &c &c, nor do I presume you will ever brew your own beer or make your own candles, therefore I sent you nothing of that sort." Unlike New England town women, Chesapeake women could not purchase locally made beer, yeast, or malt. A study of craftsmen in one Chesapeake county from 1690 to 1760 found over eight hundred craftsmen, but no brewers. Indeed, the first advertisement for commercially available beer in either the *Virginia Gazette* or the *Maryland Gazette* did not appear until 1746.[23]

Even the most basic ingredients of ale, malted barley, was uncommon in the Chesapeake. To make malt, the brewer needed a large vessel for soaking the grain, a ladle to draw off the water, a shovel or fork to turn the grain, and a device such as a kiln for drying the grain, as well as the oats or barley. Once the grain began to produce sugars, it was dried, at which point it became malt. The brewer then ground the malt and mixed it with boiling water in a mash tun or other vessel until the mash reached a pudding-like consistency. Next the brewer drew off the wort, or liquor, mixed it with yeast and herbs, and left it to ferment, producing a strong beer ready for drinking in a couple of days. To brew small ale, the brewer made a second brewing from the spent mash of the first brewing.[24]

Chesapeake women could not make even this simple, unhopped ale because they did not sow oats or barley, which were more labor-intensive than corn. "The richer sort generally brew their small-beer with malt, which they have from England," wrote Robert Beverley in his 1705 history of Virginia, "but for want of the convenience of malt-houses, the inhabitants take no care to sow it," so small-planter households had no malt from which to brew. Malting itself required labor, even if one wanted to try malting corn. "Few of them are come to malting their corn, of any kind, at which I was much surprized," commented one traveler, "as even the Indian grain, as I have found experimentally will produce an wholesome and generous liquor." Chesapeake colonists found it easier to rely on cider than to make ale.[25]

It was not only with regard to alcoholic beverages that the early Chesapeake was an unusual English province. Alcoholic beverage production was only one realm in which Chesapeake colonists reduced unnecessary labor. For example, colonists quickly learned to fence their vegetables rather than their animals, to reduce the amount of necessary fencing and labor. Despite numerous attempts on the part of the Chesapeake legislatures to create English-style capitals and towns, Chesapeake colonists did not do so because they preferred to focus their efforts on raising tobacco for sale. Colonists kept English laws and governance but simplified them to meet their needs. In England, for example, provincial assemblies and courts of appeals were separate affairs, but in the Chesapeake they were frequently the same. Colonists continued to celebrate traditional English holidays, but they added celebrations for the anniversaries of successes in battles with Native Americans.[26]

Chesapeake colonists adapted to the New World with their alcoholic drinks as well, making beverages that were unfamiliar to men and women in England or elsewhere in the Atlantic world. In particular, persimmon "beer" was not an English tradition, since persimmons were unknown in Europe. A Frenchman touring Virginia in 1686 commented upon seeing a persimmon that "there are quantities of a kind of tree that bears fruit as large as apples. Its flavor is excellent and it is pleasant to see." In 1736, an English traveler in the Chesapeake recorded that "we gathered a fruit, in our route, called a parsimon, of a very delicious taste, not unlike a medlar, tho' somewhat larger: I take it to be a very cooling fruit, and the settlers make use of prodigious quantities to sweeten a beer . . . which is vastly wholesome." In 1735, Virginian Peter Collinson sent some persimmon seeds to his friend in England, a famous horticulturalist who had never seen them. Even as late as 1777, British travelers in the Chesapeake were unfamiliar with persimmon beer. One noted that "there are amazing quantities of persimons all along the road. The Virginians make beer of them." West Turner, a former slave, remembered: "We made persimmon beer, too. Jest struck our persimmons in a keg with two or three gallons of water and sweet potato peelings and some hunks of corn bread and left it there until it began to work." Former slave Sarah Fitzpatrick remembered that while "white fo'ks" in the nineteenth century "dey made all de wine dey needed f'om blackberry's, grapes an' dey made plenny apple cider out'ta hoss-apples," slaves "made big cases uv simmon beer fer dey 'selves an' dey made good simmon bread too."[27]

The corn-based alcoholic drinks with which some Chesapeake households supplemented their fruit alcohols were also unfamiliar to the English. One of the most common of such brews was made of coarsely crushed grain mixed with water, which made a sour, fermented beverage. Another was a mix of water and corn made the same way, which formed a sharp, sweet, fermented drink. Some women added honey to these mixtures to create a type of mead. "We had a hodge podge of breakfast," griped one traveler of a meal in Virginia, which included "only one gallon of corn toddie," even though it had cost him two dollars. "The poorer sort brew their beer with molasses and bran," commented Robert Beverley, "with *Indian* corn . . . ; with persimmons dried in cakes, and baked; with potatoes; with the green stalks of *Indian corn* cut small, and bruised; with pompions [pumpkins]."[28]

While Chesapeake men appreciated women's traditional and New World drinks in the Chesapeake, making alcohol did not give women any special status or power in the seventeenth or early eighteenth centuries, mostly because women did not usually earn money by selling their drink. In the Chesapeake there were no markets for women to sell their cider to and few neighbors close by. Besides, women could not make enough cider to keep their households in drink year-round, leaving them little surplus to sell. What households did for drink when women's cider ran out, is the subject of the next chapter.

CHAPTER THREE

"This Drink Cannot Be Kept During the Summer"

Large Planters, Science, and Community Networks in the Early Eighteenth Century

When small-planter households in the early eighteenth century ran out of cider, they had a unique problem. Unlike people in England, New England, or the Middle Colonies, Chesapeake colonists could not purchase alcohol at local markets or commercial distilleries or breweries. Chesapeake residents did not develop public markets until the late eighteenth century, distilleries until the nineteenth century, or breweries until the twentieth century. Most could not import foodstuffs from England or the West Indies because they had neither the connections nor the credit or reputation to do so. The irregularity of shipments and the difficulties inherent in shipping beverages made it impractical for even the upper sort to import much liquor. Instead, Chesapeake colonists depended on the productions of nearby large plantations that made enormous amounts of cider. When small-planter households could not make cider or when they ran out, they turned to large planters for their surplus stocks of cider and distilled apple and peach brandy—at a price.

"Small planters" were those whose estates were worth less than £100 when they died; such households comprised almost 77 percent of estates. Large-planter households were those worth more than £250; middling planter estates were worth between £100 and £250. The large planters were the 3 to 10 percent of all white heads of households who rode in coaches, built mansions, owned more than twenty slaves, controlled the region's political institutions, sold their tobacco on consignment directly to merchants in England, and imitated the dress, architecture, education, leisure activities, and market consumption of the English gentry. It was this minority who imported, albeit belatedly, the technology, English connections, and knowledge to preserve alcoholic beverages year-round through the expensive and complex process of early modern distilling.[1]

The limited evidence that remains indicates that it was men who managed the larger cidering and distilling operations; their wives and other women on the plantations were far less likely to have the physical freedom, connections, literacy, financial credit, or time to handle a large distilling endeavor. While women in small-planter households made cider just as they made cheese, candles, and other household supplies, it appears that on the large plantations slaves of both sexes made the alcohol.

Small-planter households could not afford the technology or labor to be self-sufficient in the early eighteenth century and thus depended on large planters for alcohol when their own production was insufficient. The neighborhood choreography of large planters selling their plantation-produced distilled liquors to lower- and middle-sort colonists within a three-to-five-mile radius reflected the necessity of community reliance as well as the graduated nature of the region's ties with the Atlantic world. While the largest planters imported goods from Europe and slaves from Africa and maintained connections with European merchants, small planters did not. Even large planters struggled to keep up with European science and technology, and small planters were further behind. In the early eighteenth century, small-planter households found their opportunities to interact with the Atlantic world limited and focused instead on cultivating commercial relationships with large planters.

The main reason that small-planter households faced alcohol shortages was that apples were in season only from late summer to early winter. One Virginia traveler recorded that in July, before the apples had ripened, colonists "have but little cider, no beer, and the water . . . is reckoned very pernicious to the bowels when unmixed with spirit." Another visitor concluded that around "the meaner sort you find little else but water amongst, when their cider is spent."[2]

Even during the apple season, women frequently found their attempts at cidering thwarted. In September of 1736, for example, the *Virginia Gazette* warned that "there will be a great scarcity of cider, the apple-orchards having generally failed" because of a drought. Cattle killed fruit trees by stripping their bark. "Many that have good orchards," Robert Beverley critiqued in 1714, "expose the trees to be torn, and barked by the cat[t]le." Cattle injured orchard trees frequently enough for the Virginia Assembly to pass laws imposing punishments for letting one's cattle damage another's trees. A first or second offence incurred a fine of two hundred or four hundred pounds of tobacco, respectively, payable to the orchard owner. A cow or horse that injured an orchard a third time risked death.[3]

A variety of other natural threats made cidering uncertain. Unfenced pigs harmed orchards: "The home orchard much injured by the sows and pigs," recorded an irate Landon Carter. Mice, insects, and caterpillars regularly destroyed orchards. William Byrd II concluded in 1729 that his orchard fruit was devastated by worms; Thomas Jefferson declared later in the century that the "weavil of Virginia" had demolished his fruit. Frost killed fruit as well: "This changeable weather," one writer noted, "is a hindrance to the growth of fruit-trees in the region, where the warm spring tempts out the blooms very early and late malicious frosts are often very damaging." "The frost," observed another traveler to Virginia, "has been much more severe and fatal here than in the northern colonies. The peaches . . . are wholly destroyed, and these were the choicest expectation of some, who think brandy their most valuable commodity." "I should have never done," concluded another essayist, "were I to recount to you all the inconveniences . . . the trees of our orchards . . . are exposed to. . . . These calamities remind us of our precarious situation." Finally, men whose households lacked women did not have the time to cider, even when apples were abundant.[4]

Small-planter households had limited options for cider replacement. Until the late eighteenth century, Chesapeake colonists had few towns, marketplaces, or merchants, even in Williamsburg and Baltimore. When large planter Robert Carter moved to Williamsburg in 1761, he found that he had to purchase a small farm nearby to raise food for his family, since there was no marketplace to supply them. Nor could colonists trade for alcohol with Chesapeake Indians. While many natives in the Americas made alcoholic beverages, the Pamukey and other tribes in the Chesapeake generally did not. Chesapeake colonists could not easily purchase alcohol sold by colonists in New England and the Middle Colonies, since there was little coastwise shipping until the latter half of the eighteenth century. When John Custis of Virginia wanted some plant samples from a Pennsylvanian in 1739, he complained that "it is more than a hundred to one if ever he meets with an opportunity[,] no vessels ever coming from that place hither." Virginians saw only three or four ships a year from New England ports, and ships from Pennsylvania and the lower South docked even less frequently. Ships from England were more common; by the end of the seventeenth century, about 150 ships landed in Virginia from England annually. Colonists in the Chesapeake could not expect to get goods from any place other than England on a regular basis.[5]

Even the alcohol that Chesapeake colonists imported from England was limited. Records listing the products that Virginia and Maryland colonists imported from London from 1724 to 1774 reveal surprisingly little alcohol. The ledgers indicate that the Chesapeake region produced most of the alcohol that it needed. From 1724 to 1774, ale, beer, and cider together averaged only .16 percent of the value of the goods that Chesapeake colonists imported. If the 1774 to 1775 records are excluded from the analysis to avoid the influence of the 1770s non-importation agreements, then the value of ale, beer, and cider imports averaged just .20 percent of Chesapeake colonists' spending on imports. Wine imports were slightly higher; they averaged .32 percent of all purchases. In total, less than one-half of one percent of colonists' expenditures on all imports between 1724 and 1774 went to wine, cider, beer, and ale combined.[6]

Large planters tried to import liquor, particularly types that they could not make on their plantations. Robert Bennett thanked his "loving brother," Edward Bennett, for sending "19 buttes of excellent good wines." Robert Carter wrote to his usual English merchant that if he shipped him "18 dozen of good white wine . . . [and] 20 dozen of right good ale it would not be amiss." Still imports were limited.[7]

Large planters imported little alcohol because importing liquids was unreliable. The cost of imported alcohol was particularly discouraging when the product was not of the anticipated quality or ultimately proved undrinkable. Richard Corbin recorded angrily that the beer and other goods he had ordered from England "are not equal in price" to what he had paid for them. Two months after Colonel James Gordon "bought fifteen gallons rum, at 6/ [shillings] — the greatest price I ever gave," he decided to try to distill his own, "which I believe will answer very well, as liquor is so very dear." Sometimes alcohol arrived so foul that it was impotable. In 1729, Robert Cary complained to his supplier that "you sent me a hogshead of beer as it is called bought of J. R. Pycroft. Such trash was never before brewed. I offered it my negroes, and not one would drink it; the worst molasses rum that ever was brewed was cordial to it; I flung it every drop away." "It is not at all a pleasant wine," John Custis wrote to English merchant Jonathan Day in 1736; "I cannot drink one drop of it." "The beer you sent me," he recorded on another occasion, "is the [most] rascally stuff I ever tasted sour [and] poor . . . if I could have had good beer from England should always had it from thence, but I find [importation] will not do . . . it is a hard case a man cannot have reliable usage for his money." Robert Carter complained that the brandy he imported from James Arbuckle was "such I never met with before that bore the name of French brandy. The person you bought it of certainly changed it or mixed it after you had tasted it." Custis requested that his supplier watch his tradesmen so that they would "not play such villainous pranks as they do; they send more trash . . . some things are sent that I never desired or wanted, and others left out which I much want. . . . I have heard the like complaints from several people." Custis's letters reveal constant difficulties with importing alcoholic beverages, from shrinking bottle sizes that "every [time] I have them from you they are made to hold less and less," to half-filled bottles, which caused him to ask his suppliers to "see that they hold full quarts." Landon Carter's liquor leaked because, although he was "charged with waxing the bottles, [but] that is only a black seal at the top of the corks." John Custis discovered that he had been overcharged for iron hoops on his wine pipes that burst and spilled his wine. More of Byrd's bottles broke "partly owing to the careless way of packing and partly to your masters tumbling them ashore at Hampton and tossing them into a warehouse." In one case, Byrd never would have found his bottles had they not been mistakenly loaded upon a ship that stopped in his area by chance.[8]

Many of the hazards of importation were practically unavoidable. Hoops broke on wine pipes and barrels, releasing the liquor into the ship. "The hoops flew off one of the casks," lamented Robert Cary, destroying a cask that was "excessive dear." Lord Botetourt ordered four pipes of wine, only to see them arrive in disarray: "Casks are crazy and it is apprehended if they were to be moved before some cooperage, that the wine would leak and be lost." Storms, shipwrecks, and piracy destroyed imports of all kinds. William Byrd wrote that Captain Hunt brought him seventy-seven empty bottles "from the vessel that had been taken [by pirates] and lost my cider." Imports also might fail because of mislabeling or malfeasance by merchants who knew that the region's planters were powerless to sue.[9]

Eighteenth-century plantation records indicate that large planters made enough alcohol on their plantations to sell the surplus to nearby tenants, artisans, and small planters. The earliest known surviving Chesapeake plantation ledger shows something of how this neighborhood trade worked. The ledger is from the Littletown plantation of James Bray III, a 1,280-acre tract about five miles southeast of what is today Williamsburg, Virginia. Bray's secretary, Carter Burwell, kept a ledger for the Littletown plantation from 1736 to 1746 that has survived to reveal that while the Bray plantation was on the small side for a large planter, it was effectively a small town: Bray's neighbors, tenants, employees, and slaves came to him for brandy, rum, cider, shoes, hats, wheat, cash, corn, tobacco, chicken coops, chickens, linen, pistols, butter, and mutton, among other goods. By 1740, Bray not only grew tobacco but also had built a commercial gristmill to grind corn and wheat, a tannery, a lumber manufactory, and a cooperage, as well as brickmaking and shoemaking facilities. He then sold the surplus production from these operations to the households around him as well as to his tenants and laborers.[10]

The Brays were typical of Virginia's large planters. James Bray Sr. married Mourning Glenn Pettus in the 1690s and gained title to her property, Middle Plantation. Bray was a member of the Governor's Council, and his descendents became burgesses, vestrymen, and county officers. Their social and political standing entitled them to be called "gentlemen" and to attain prestigious militia ranks such as colonel, captain, and major. Bray bequeathed the Littletown land to his son, James Bray Jr., who left the land to his daughter, Elizabeth Allen, until her son, James Bray III came of age, probably in 1736. The Littletown plantation house was a two-story, 4,500-square-foot dwelling, where James Bray III lived with his wife, Frances Thacker, until he died. (Littletown plantation then reverted to Elizabeth, who was still living.) James Bray III owned at least eighty slaves, three indentured servants, and five hundred head of stock. Most symbolically, Bray owned a money scale and steelyard, or balance beam scale, to weigh and balance accounts. Just as a ring of keys and a pocket were the signs of the housewife's labor in dispensing foodstuffs from cupboards, so the money scale and steelyards were the symbol of the planter-merchant who weighed coins and crops.[11]

The Littletown records have limitations. The plantation ledger is so tangled and incomplete that performing a traditional analysis on it is impossible. Researchers cannot even determine the total number of Bray's customers, because he appears to have called the same people by a variety of names. Yet the account book offers a glimpse of the neighborhood economy in the Chesapeake in the early eighteenth century. It reveals that Bray sold goods to all layers of society, from "My Lady Randolph" or the attorney-general of Virginia to Williamsburg craftsmen and Bray's own overseers and tenants. More of Bray's recorded transactions were for loans than for sales of merchandise. Small planters, artisans, and laborers could pay for local goods and services outside of Littletown, such as the Williamsburg printing office, because Bray and other large planters loaned them the money.[12]

The primary purpose of Littletown and other Chesapeake plantations was to raise and sell tobacco. It is likely that Bray sold much of his tobacco to Robert and John Lidderdale. Robert Lidderdale was a merchant in London; John handled their business interests in Williamsburg. Through the Lidderdales, Bray obtained the goods that he wanted from Europe, which they supplied to him in order to gain his sizeable tobacco crop.

The Lidderdales did not deal with small planters. In the first half of the eighteenth century, small planters depended on Bray and other large planters like him not only to provide the goods they needed but also to sell their tobacco. Nathaniel Overstreet, for example, came to Bray several times in 1740 for a pair of shoes, a bottle of imported beer, some cloth, and several loans. Overstreet then gave Bray his tobacco crop as partial payment for the goods and promised to pay him an additional five pounds, which included the fee that Bray charged for marketing Overstreet's tobacco. Bray added Overstreet's tobacco to his own and sold it to his connections, the Lidderdales. The Lidderdales shipped the consignment to England and sold the tobacco to European buyers.[13]

Such transactions, with the large planter acting as a middleman, were common. For instance, when members of the Page family could not obtain wine from England, they turned to large planter John Carter of Corotoman, who contacted his merchant connection in London, Micajah Perry. "Mrs. Page desires that you will order two pipes of wine from the Madeira for her house," Carter explained, and requested that Perry "let her have the best that the island affords." Likewise, when a local doctor died, his executors asked Carter to market "the small crop belong[ing] to Dr. Nicholas's estate."[14]

Sometimes planter-merchants built special outbuildings on their plantations expressly to sell goods to small planters and laborers. Planter-merchant William Gordon, for example, owned thirty-one slaves and seven white servants, and he sold pewter plates, spoons, shoemakers' tools, earthenware, eyeglasses, buttons, material for curtains, Madeira wine, French brandy, and flower pots from a small shop on his plantation. Other planter-merchants sold their neighbors butter, cheese, cloth, thread, scissors, and tankards through their plantation stores. Most of these accounts were between men. Although Bray hired someone to keep his books and run his store, he may have met with his customers himself as well to engage in patronage, seek votes, and exchange news.[15]

Robert Wormeley Carter also sold plantation-produced and imported alcoholic beverages to his neighbors, including his son, his doctor, large planters and small planters close by, and a nearby tavernkeeper. For example, when Carter sent six gallons of peach brandy to his son, he recorded that it cost him £11, because he expected to be repaid. His son paid him for the brandy in October 1774. Carter traded alcoholic drinks with his neighbors when it suited him, such as when he purchased twenty-six gallons of cider from his family doctor, a large planter in his own right. Later Carter sent two dozen bottles of his imported Madeira wine to Mr. Tayloe "in return for the like quantity" that Tayloe had given Carter earlier. In July of 1792, Carter exchanged home-produced brandy for oysters from planter James May. When Carter traveled, he purchased alcohol himself as when he left on a four-day visit "among the freeholders" in July of 1769. On that visit he purchased two quarts of rum from Colonel Fantleroy and paid James Gordon for an additional two quarts of rum to distribute to his political supporters. A week later, when Carter went to a muster at Reid's Field, he bought two and a half gallons of rum from Will Miskell to give to Carter's militia. Carter bought a bowl of grog from tavernkeeper John Howe while away from home in December of 1787, and he noted carefully that it contained less than the seven gills of brandy he had been promised. In August of 1790, Carter paid Howe for grog and noted that Howe owed him another gill of rum. Such accounts suggest just how regularly large planters made surplus alcohol to trade and sell with nearby colonists.[16]

"There is little opportunity to sell eatables, except in harbors and in the inns," one traveler recorded. With no alternatives, small planters depended on large planters for goods, including alcohol when their own production was not enough. Artisans, tenants, and small planters depended on Bray for surplus meat, butter, and corn. One historian calculated that Bray sold over 4,700 pounds of adult beef in small lots to neighboring small-planter households and tenants between 1736 and 1746. He sold approximately 550 pounds of veal, 12,000 pounds and 4½ barrels of pork, over one hundred pounds of shoat (young hog), two hogs, two "roasting pigs," and eighty-four pounds of bacon to his tenants and employees. Bray also sold mutton, plantation-produced butter, and corn; and for a fee or a cut, he would have his laborers mill the corn that his neighbors brought to his three-man mill.[17]

Large planters like Bray also supplied food to local institutions. Carter Burwell, for example, annually sold 600 pounds of beef, between 500 and 800 bushels of milled wheat, and other products to the Governor and the College of William and Mary. He also supplied local tavernkeepers, selling 100 bushels of wheat to John Doncastle of Williamsburg in 1755. Tavernkeepers Anthony Hay, Thomas Penman, and Christianna Campbell all took 25 bushels of wheat from Burwell at that time too. Susannah Allen bought cider for her Williamsburg tavern from William Byrd's plantation, John Major sold distilled liquor to tavernkeepers from at least 1772 to 1794, when he decided to open his own tavern. Henry Weatherburn and Ann Pattison both bought foodstuffs, pipes of cider, and firewood for their respective taverns from Bray. Mrs. Flemins, a schoolmistress, purchased firewood from Bray for her school.[18]

Neighbors depended on large planters like Bray for cloth or for raw materials, shoes, bricks, and wood. He sold cords of wood, timber trees, and products from his cooperage, including planking, lathing, clapboards, scantling, siding, heading, fence rails, fence posts, framing, and coffins. In addition, the plantation's coopers made pails, ladles, hogsheads, casks, pipes, barrels, tubs, and hoops, which Bray sold to his customers.[19]

And Bray sold cider. The advantages of Bray's plantation included large orchards, a cellar with a sump drain, and a cooper who made cider casks. During the ten years that Burwell recorded Littletown's sales, Bray sold at least 1,009 gallons of cider to his neighbors and at least 120 quarts of cider to his tenants and overseers. Bray or the plantation's laborers distilled some of his cider, selling at least 345 quarts of apple brandy in 1743 and 1744 alone. Bray also sold a little "Syder

Royal," an import of English cider mixed with brandy to help it survive the voyage. When Bray died in 1744, his plantation store held fifty gallons of brandy, one cask and more than 211 gallons of cider, and seventeen "dozen," which probably referred to the number of bottles of cider royal on hand for sale to neighbors. In addition to cider and apple brandy, Bray also sold his neighbors small amounts of wine, rum, tea, sugar, molasses, coffee, snuff, chocolate, oranges, salt, and cheese, all of which he had imported.[20]

No matter who actually produced alcohol on a large plantation, the records inevitably were kept by men. Most Chesapeake women were illiterate and innumerate, and all plantation records were typically kept by a male plantation secretary. For example, Robert Wormeley Carter had his male overseer, William Petty, keep accounts of brandy sales. Burwell kept records of Littletown's sales of butter, although the butter was likely made by the overseer's wife and the plantation's enslaved women. Landon Carter also recorded sales and purchases of butter, even though dairying was women's work. Anne Pattison kept a tavern in Williamsburg for at least five years, but she had her male barkeep, James Lebe, keep an account book for her. Another tavern owner, James Southall, placed an advertisement in the *Virginia Gazette* in 1766: he sought "a young man qualified to act as barkeeper, that can write a tolerable hand, and understand something of accounts."[21]

It is likely that men oversaw cidering at Littletown and the few other large plantations in the region, although evidence is sparse. The large orchards needed for large-scale production of cidering extended far from the house, putting them outside of the usual English custom of women working within households and garden gates. Women could not travel to larger orchards because they needed to be near home to nurse children and cook meals. It was easier for men to travel to court, town, other plantations, and to tenants to manage the sale and trade of cider, as well as to a bookstore, ship captain, or another large plantation to obtain a book on cidering or orcharding. Finally, married women were *femes covert* and thus not allowed to own property or sue in court. In small-planter households where production was limited, it made sense for women to make cider; in large-planter households where production was more complex, it was logical for men to take over this task.

Only a few extant sources acknowledge that it was slaves, often referred to euphemistically as "people," who performed much of the labor of producing alcohol, and most of these documents are from the latter part of the eighteenth century. For instance, Phillip Vickers Fithian noted in 1774 that at a Virginia plantation the "people were shaking the trees to prepare the peaches for brandy." Robert Wormeley Carter recorded that he purchased apples from a slave. Landon Carter had slaves cart fruit and beat it into cider, and William Byrd also watched "my people" make alcoholic beverages. John Mason's father, owner of the Gunston Hall plantation, relied on a slave for distilling at times. Thomas Jefferson recorded his slave's advice about the amount of cider to be expected from each bushel of apples: "George says that when in a proper state . . . they ought to make 3 gall[on]s to the bushel, as he knows from having often measured both." In 1773, Norfolk passed a law prohibiting "mulattoes or negroes bound or free from selling . . . spirituous liquors." George Washington's weekly farm reports in the late eighteenth century list the number of slaves assigned to each agricultural task, including alcohol production. William Byrd recorded early in the eighteenth century that "I took a walk to see my people plant peach trees." Byrd also wrote that "I caused Bannister to draw off a hogshead of cider." A typical entry in planter James Gordon's diary reads only: "Apples now down. We expect to make 500 gals. cider, or thereabouts." Although few, if any, other references to Chesapeake slaves' production of alcohol remain, even these scant allusions indicate that slaves did much of the work of making alcohol on the few large plantations.[22]

Although sources do not indicate the gender of the slaves who produced alcohol on large plantations, it is possible to advance the cautious conclusion that they were both men and women. English planters with rum operations in the Caribbean used both male and female slaves to make alcoholic beverages, with female slaves performing simpler tasks and men more complex ones that required training. Large planters in the Chesapeake appear to have used both male and female slaves to make alcoholic beverages as well, which would explain why their accounts do not mention the sex of the slaves making alcohol.[23]

Large planters engaged white men to oversee the slaves who were making alcohol. George Washington's nephew sent Washington a "gross of crab cider" with apologies, explaining that it was "not so good as I could wish, from the management of my cider last fall being left entirely to the negroes, from the loss of both my overseers." At George Washington's Mount Vernon plantation a few years later, a guest recorded that the "lands are divided into four farms with a number of blacks attached to each and a black overseer over them. The whole is under the supervision of Mr. Anderson, a Scottish farmer . . . under the supervision of the son of Mr. Anderson, they distill up to 12 thousand gallons at year." Washington relied on white overseers to manage his slaves' cidering as well. "Bishup has . . . made two casks of cider," noted a letter to Washington about the overseer of his Muddy Hole farm. Washington had a white overseer manage cidering on his River Farm on Clifton's Neck as well, where "Cleveland . . . made four casks of cider." The employment contract between George Washington and overseer Nelson Kelly elaborated on this role: Kelly was, "in case the said George should judge it expedient, to beat the apples which may be found upon the plantations into cider making as much as he can thereof."[24]

Large planters like Bray produced far more alcohol than they needed for their families and laborers alone, and they sold that surplus to their neighbors, something that scholars have not realized. Even in the seventeenth century, large planters made vast amounts of cider and perry. In 1675, Peter Marsh pledged to pay James Minge 120 gallons of surplus cider. In 1676, Thomas Glover wrote in *An Account of Virginia* that "there are few planters but that have fair and large orchards, some whereof have twelve hundreds trees and upward, bearing all sorts of English apples . . . of which they make great store of cider. Here are likewise great peach-orchards, which bear such an infinite quantity of peaches, that at some plantations they beat down to the hogs forty bushels in a year." Shortly thereafter, Alexander Moore bequeathed twenty gallons of raw cider and 130 gallons of boiled cider in his will. In the late seventeenth century, Richard Bennet made at least twenty butts (approximately 108 gallons) of cider annually, and Richard Kinsman produced forty to fifty gallons of perry annually. Richard Allen of Surry County wrote to John Owen in 1704 that Owen could fetch twenty hogsheads of surplus cider from Allen in September.[25]

Sheer production capacities suggest sales, even for those planters who did not leave alcohol records. For example, Colonel William Fitzhugh had 2,500 apple trees, indicating a capacity of at least 9,600 gallons of cider annually, on a plantation served by twenty-nine to fifty-one free men and slaves. Even when the Fitzhugh plantation housed fifty-one people, that population required 2,320 gallons of cider at most, leaving at least 7,280 surplus gallons of cider. One year, Fitzhugh sold this surplus for "at least 14,000 lbs." of tobacco.[26]

Planters like Fitzhugh were probably even more productive than this estimate suggests. In Pennsylvania, for instance, Henry Wynkoop produced 4,480 gallons of cider from only two hundred trees. Robert "King" Carter of Corotoman plantation, Virginia, recorded in 1722 that his slaves had made over twenty butts of cider, or at least 2,880 gallons, from August 15 to August 29 alone. Colonel James Gordon's plantation produced 1,500 gallons of cider just at the start of the cidering season in 1763. William Cabell's Amherst County, Virginia plantation fermented 3,000 gallons of cider and fifty hogsheads (at least 2,400 gallons) of peach mobby annually. Traveler Josiah Quincy, Jr. recorded seeing peach orchards in Virginia of at least thirty acres, cultivated in order to make peach brandy, "a favorite liquor." In 1765, Carter Burwell sold forty-eight hogsheads (at least 2,300 gallons) of cider from his estate, Carter's Grove, for a little over £83. Robert Wormeley Carter made alcoholic beverages as well. In 1768 he planted at least 1,200 peach trees intended for peach brandy, and experimented with inoculating his fruit trees. Thomas Mallory asked a merchant to send him "242 yards of wide hair clothe fit for pressing cider" to support extensive cidering. Robert Burwell advertised his plantation for sale in 1771 with "at least 1,700 apple trees and a great variety of other fruits." Landon Carter recorded that in 1778, because of extensive rain, he was setting up "only" nine ninety-gallon casks for cider.[27]

Cidering allowed large planters to make use of poor, sandy, or thin land that was unfit for tobacco. Traveler Isaac Weld noted this on his trip to the sandy land of the Dismal Swamp. "The soil is so poor that but very little corn or grain is raised," Weld observed, "it answers well however for peach orchards, which are found to be very profitable. From the peaches they make brandy." By 1680, surpluses of alcoholic drink and foodstuffs provided, on average, 20 percent of large planters' income.[28]

The large planters' surplus was evident in advertisements of properties for sale. Planters advertising land in the *Virginia Gazette* announced how many gallons of cider could be made from the orchards on the property. John Ralls advertised that his plantation, up for sale, made 7,000 to 8,000 gallons of cider annually. William Claiborne advertised that his three orchards yielded 10,000 gallons of cider per year. In a typical advertisement, Drury Stith listed his plantation for sale in 1737. He announced that his property had "six hundred acres of land, with a good apple orchard, of choice grafted fruit, a good dwelling house, 25 by 30, with a brick chimney and a cellar." Some cidering operations were large enough to require their own buildings. Large planter Richard King advertised in 1727 that his property had a separate still house. Landon Carter, Isaac Gibson, John Mercer, and Robert Wormley Carter all advertised that their properties had specialized buildings for plantation breweries and distilleries.[29]

Large planters could produce alcoholic drink surpluses year-round when others could not because of a variety of technological and labor advantages. The first was their ability to graft and propagate orchards. Apple trees were not indigenous to North America, and growing orchards required time and labor. Apple trees grown from seeds produced sour, inedible fruit and unappealing cider. For sweeter fruit and palatable cider, planters had to graft or "inoculate" the trees individually and monitor them through occasional re-graftings for ten years or so until the trees reached maturity. Inoculating was expensive and required constant attention. Robert Wormeley Carter recorded in 1780 that he paid a visiting gardener $10 to inoculate just "4 peach trees & 1. mulberry; with apricot and English mulberries." George Washington hoped to employ a gardener who understood "something of fruit trees and could graft and inoculate" for a permanent post. He hired one who, "to show his cunning," grafted peaches onto plumb scions. Richard Henry Lee hired a permanent gardener to graft fruit as well, reminding him "to keep a nursery well supplied with good young graftings of all kinds of fruit trees, particularly of the choicest and best apples both for cider and eating." William Fitzhugh concluded that it took at least seven years to prepare and graft an orchard and warned that "without a constant care and continual residence there upon, the labor and care of seven years is destroyed in as many hours." The cost and "constant care" of grafted fruit explains Robert Beverly's 1705 criticism that "yet there are very few planters that graft at all, and much fewer that take any care to get choice fruits." Only the wealthy could afford to graft their orchards for improved cider and edible fruit.[30]

Large planters also had a great variety of cuttings, which allowed them to cultivate more robust orchards. Orchards with a variety of fruit crops were less susceptible to devastation by disease. Large planters shared their cuttings with each other: Landon Carter obtained several thousand peach trees from his wealthy neighbor, Joseph Ball; and when George Washington wanted to try a new type of fruit, he acquired his scions from Colonel George Mason. Thomas Jefferson developed better peach trees through exchanges with Mason as well. John Custis of Williamsburg established three varieties of English peaches in his garden, using peach pits and young trees sent to him by Peter Collinson, and he began distributing scions of these peaches to his friends once the trees were large enough. William Byrd wrote a friend in England who had sent him cuttings: "Thank you kindly for the cuttings, you sent me, of choice apples, & pears; which I immediately, caused to be grafted." Wealthy planters also experimented with the new breeds of trees found on the western lands they had purchased and whose explorations they financed. George Washington told one agent he sent to explore his new Ohio River lands that "if you could get peach, or any other kind of fruit stones, or apple seeds, it would not be amiss to engage them to carry out with you."[31]

Being wealthy enough to own a large orchard was not so much a matter of buying land as of marshaling the labor required to plant trees. Robert Wormeley Carter noted that on February 20, 1768, he had his Hickory Thicket plantation overseer and slaves plant 623 peach trees. One week later, Carter paid the overseer to plant an additional 600 peach trees purchased from another large planter, Peter Northerns. Carter also had the overseer move fruit trees from another of Carter's quarters to the Hickory Thicket plantation. By the end of February of that year, the overseer and his crew had planted or transplanted "in all 1500" fruit trees to the Hickory Thicket plantation.[32]

Grafting was laborious, but necessary, to grow edible fruit. *The Expert Gardener, or, A Treatise Containing Certaine Necessary, Secret, and Ordinary Knowledges in Grafting and Gardening* (London: printed by William Hunt, 1654), 33. Image used with permission of The Rare Book and Manuscript Library of the University of Illinois, Urbana-Champaign.

Orchards required that men place trees properly as well as weed, fence, and generally look after them. Husbandry advice author John Worlidge reminded planters that "there is required some judgment from the husbandman in placing each tree or plant in the proper soil it most delights in," and he advised readers to choose "good warm light" soil for "cider-fruit" because the "heavier, colder, and moister" land "is not so good, the cider being not so clear." In 1743, Joseph Ball warned his nephew in Virginia that he must "chop over" the broom grass regularly, "else the grass will kill the trees." Ball instructed his nephew to build "a strong crotcy fence" around the trees "to keep cattle, and horses, from tearing and barking" and killing the orchard trees. Someone had to monitor the trees for sickness and to remove or cure those that were ill before the disease spread. Curing ill trees involved reading the literature concerning orchards. William Byrd II, for example, carefully copied into his commonplace book instructions for caring for trees with what the author called "cankers." "When a fruit tree is cankerd rub off the canker," Byrd noted, "and then apply tar and greese to it and twill recover itself again."[33]

Planters had to move orchards and transplant seedlings regularly. Tobacco was hard on the soil, so planters rotated their crops, alternating between grains, tobacco, and orchards. The amount of labor necessary to rotate an orchard was enormous. Joseph Ball told his nephew that moving the peach orchard would require so much labor that the plantation would not be able to complete its other goals: the "peach trees must all be moved in early next spring, let what will go undone," Ball instructed. The nephew's plantation labor force alone would not be enough, Ball warned and, "they must have some help from the other plantations with carts and men."[34]

Large planters could afford to purchase apples from other regions if disasters, such as orchard-destroying storms or hungry cicadas struck their plantations. A bushel of apples cost between one and two shillings in Virginia throughout the eighteenth century, with the cost rising in the winter as apples became less available and small households ran out of cider. At one to two shillings per bushel, apples were a costly purchase for small planter households: a pair of shoes cost on average one shilling and a hog, six shillings. It took twenty to thirty bushels of apples, at a total cost of £1.17, to make a hogshead of cider in an age when the average small planter made only £11 annually.

The most notable distinction between large- and small-scale cidering was the use of the industrial cider press. In England in the late seventeenth century, men raced to build better cider presses called "ingenios" to speed the process of pressing the juice out of apples. Inventors announced their new devices in books. John Worlidge advertised in 1678 that he had added to his book on cidering "the true way of making the Ingenio for the grinding of apples . . . so useful a machine." Worlidge included several drawings of the horse-powered machine, which pushed apples between rollers, sometimes toothed, to crush the fruit. Henry Allen advertised in Worlidge's book that he had devised improved screwpresses for pressing cider, with "both screw and nut being of cast iron, so tempered, that they shall never fail." Allen promised that his screwpresses "operate more effectually, and with much less labor than the great wooden skrew-presses." The ciderer then strained the pressed juice through cloth.[35]

Chesapeake colonists found it difficult to obtain the new cidering inventions. The first challenge was to learn what was available. In 1694 William Fitzhugh wrote to his English merchant connection: "I desire you to send me a thing to rack hogsheads of cider, which Mr. Scott tells me they have," indicating his uncertainty that this device necessarily existed. Even an elite planter such as Fitzhugh was powerless to obtain what he wanted. "I hope if you do not rightly understand me," Fitzhugh worried, "Capt. Scott will inform you." Fitzhugh was right to be concerned. When William Beverly asked his English merchant connection to send him an "ingenio or cyder mill to turn by hand with lignum vitae roll[e]rs

The new cider presses could be large, complex, and laborious, but they offered increased production to those who could afford them. John Worlidge, *Vinetum Britannicum, or, A Treatise of Cider and Other Wines* (London: Printed for Thomas Drink, 1691), 105. Image published by permission of the British Library.

brass peggs & a screw press," it took more than a year for the cider press to arrive, at which point Beverly unhappily learned that it cost twice as much as he had anticipated. The cost was likely shocking: in 1781 one Robert Babb's "syder mill" was appraised at £900.[36]

Still, large planters obtained cider presses. When Captain Daniel Taylor died in York County in 1712, he left a "syder trough and press." The executor of a planter-merchant's estate in Maryland advertised in 1747 that he was selling pewter steelyards, scales and weights, "a good desk," "a choice parcel of cattle and hogs," and "a cyder-mill and cyder-casks." After Captain Mathew Kenner died in 1744, his estate listed "2 apple mills cribs & platforms." William Baily's estate in 1777 had "1 apple mill & press, tubs, flat forms & troughs."[37]

Even the wealthiest large planters like Robert "King" Carter sometimes passed up the improved cider presses because of their cost or used a combination of new and old methods. Carter noted in his diary in 1722 that the slaves on his Corotoman plantation were beating apples for cider by hand. Planters' use of the word "beat," as in "my people beating cider," is telling. Planters without the new technology had their laborers beat apples in troughs in the traditional fashion, with cone-shaped pounders on four-foot handles. For example, Landon Carter had his laborers beat fruit for cidering in strong wooden tubs with pounders, and traveler Ebenezer Hazard recorded in 1777 that he saw planters making peach mobby "by putting the peaches in a trough and bruising them with a pestle, so as not to break the stones. They are then thrown (thus bruised) into a hopper, or a hogshead with a hole in it (and a vessel put under it) and suffered to drip."[38]

The presses were worth the trouble of getting because of the profits that large planters could make from cidering. Records are few, but among them are notes stating that Landon Carter paid fifteen shillings for a barrel of cider in 1770 and that large planter Robert Wormeley Carter paid Colonel Brockenbrough twenty-six shillings for twenty-six gallons of cider in 1777. In 1791, Carter recorded paying for cider twice, both times at "at 6d." It is fair to conclude from scattered sources that large planters charged one-half to one shilling per gallon of cider in the eighteenth-century Chesapeake. Aged cider cost more, as did cider purchased in taverns. A planter who sold a modest 2,000 gallons of surplus cider at an average of nine pence per gallon (three-quarters of a shilling), and subtracted £5 a year in production costs, stood to make an average of £70 a year in surplus cider sales. A large planter could thus earn six times the average annual income of a small planter or laborer through surplus cider sales alone.[39]

Large-scale cider production required a cellar, cooperage, and a set of specialized tools if it was to last. Small planters typically had access to none of these. Cider spoiled and turned to vinegar quickly in the hot Chesapeake climate when the maker did not have a cool place to store it. Travelers frequently complained that Virginians drank their cider too raw and did not age it enough, not realizing that small planters did not have storage cellars. Hugh Jones noted in 1724 that Virginia cider was "excellent . . . when kept to a good age" but lamented that aging "is rarely done, the planters being good companions and guests whilst the cider lasts." William Grove criticized the way that Virginians made "good cider but will not keep it but drink [it] by pailfulls never worked [aged]." Durand de Dauphine complained that "it was cidermaking time. Everywhere we were required to drink so freely . . . the cider made me ill; I think it was too new." One traveler commented on the short life of cider expressly: "As the common man does not have good cellars," he observed, "this drink cannot be kept during the summer, but it turns sour." In contrast, large-planter William Fitzhugh had four cellars, enabling him to supply small planter households when they ran out.[40]

For planters to make and store surplus cider, they also had to have secure barrels, and plenty of them. Making barrels required skilled coopers who knew what kinds of barrels preserved cider best. An experienced cooper, always a man, might produce two barrels a day. Several advice manuals on cidering recommended storing cider in the largest barrels possible. Worlidge advised in *Vinetum Britannicum: or a Treatise on Cider* that "the larger any vessels are, the better liquors are preserved in them." He warned ciderists not to use "the vulgar round barrel" for storage, but rather "the upright vessel, whose ribs are straight, and the head about a fourth or fifth part broader than the bottom, and the height equal to the diameter of the upper part, the best form to stand in a cellar. The bung-hole of about two inches diameter, is to be on the top, with a plug of wood turned round exactly to fit into it." Planters learned in the cidering literature that they should season the barrels with apple pumice and boiling water. If a barrel was "musty," then it had to be filled with boiling water with pepper for two or three days, or un-staved, shaved, "burned" (smoked) and remade. Barrels that had previously held Canary, Malaga, or Sherry wines were thought "to advance the color and flavor of" the cider. Surplus cidering, planters learned, required a variety of vessels with specific purposes. Experts warned planters not to mix their barrels and casks. Worlidge reminded that "a good ciderist will have his vessels wherein he puts his pulp or ground fruit, wherein he presses and tuns his liquor, and wherein he makes his cider." Planters were not supposed to use these vessels for anything other than cidering. Expert literature reminded planters to have their servants check their filled barrels frequently to prevent bursting from pressure built up during fermentation. Servants were ordered to release the bung holes of the barrels regularly to allow gasses to escape for this reason.[41]

Barrels would only preserve cider for so long. Even the best barrels let in air and caused cider to spoil. Planters who wanted to make cider last longer learned how to bottle their cider. In the first half of the eighteenth century, it was difficult to obtain bottles, so most colonists stored liquids such as cider in leather bags or earthenware. Much of this earthenware was a porous redware. Americans did not successfully produce glass until 1740, and then only in small amounts. Philadelphia brewer Samuel Carpenter offered beer for sale to "those that send clean bottles with good corks," revealing that even a professional urban brewer in a port city could not get bottles.[42]

In the Chesapeake, it was even more difficult to get glass for bottling. Still, probate records indicate that the large planters used their English connections to obtain glass bottles and that they re-used them carefully to preserve their cider and distilled liquors. "I bottled a cask Nominy [Nomini plantation] cyder," Robert King Carter recorded "23 doz[en] in my own bottles." Planters who loaned bottles expected them to be returned. William Byrd noted in his journal when a servant returned a bottle that Byrd had filled with medicine for her when she was sick. "I found Jenny," he recorded, "come to bring back my bottle in which I gave her something for her indisposition."[43]

The process of bottling was specialized. One had to know when the cider could be transferred from the cask to the bottle. The first step was to mix the raw cider, or "liquor," with isinglass (a product of sturgeon bladders that attracted particles) in the barrel to settle it. Prescriptive literature disagreed on many of the finer points, but most acknowledged that if the "liquor spurts up with force" from the cask, then the bottler needed to fill the bottles and leave them uncorked until the liquor settled. Even if a bottler had the timing right, the process was still complex. A bottle with too much liquor in it might burst, one with too little liquor might allow the cork to dry out and shrink. It was necessary to check bottled cider daily, since if one bottle broke, others might too. Sometimes it was necessary to uncork all the bottles to "give vent" to the fermentation gasses and then re-cork them, or risk having all the bottles explode. Judging when the liquor was ready to be moved from the barrel to the bottle took practice. Planter Landon Carter could tell when his laborer, Billy Beale, began bottling too soon and harmed his drink. As Carter observed, Beale "bottled it off so soon; that it will never be fine."[44]

Even obtaining corks was challenging. Virginian James Carter begged English merchant John Norton, "Send me immediately 100 gross of the best viol corks. . . . I can't carry on my business for want of them." The corks would be costly. Charles Robinson, an English cork-cutter, charged £2.7 for 32 gross of corks in London. Shipping added extra fees. Experts advised that bottlers not reuse corks. "Economy in corks is very unwise," warned one, because it would "run the risk of losing the valuable article it is intended to preserve." Planters had to treat many of their corks to make them create good seals. Some planters dipped corks in a mixture of rosin, beeswax, and brick-dust and then laid them out to dry. Others preferred to dip their corks in a mixture of white wax and beef suet, repeating "the dipping till they are saturated." In a pinch, colonists could plug their bottles with cloth made stiff from sealing wax or with glass stoppers, but corks kept cider longer. After waxing, the corks still had to be tied to the bottles with copper or iron wire. A ship captain listed the cost of wiring corks for bottles of rum, beer, and wine at a half-shilling per bottle in a catalog of his expenses. Planters who could not get wire could dip the corked bottles in a mixture of resin, beeswax, and sealing wax.[45]

Once the cider or other alcohol was bottled, it had to be stored properly. To keep cider cool in the summer and warm in the winter, planters laid it in sand or sawdust in their cellars. Landon Carter noted that he was preparing to bottle and store some cider. Once the cider "fines in the cask," Carter noted, "I shall bottle, tie down the corks, and wax them; and then lay the bottles leaning buried 4 inches in dry sand, according to Miller's *Dictionary* and hope to get a pleasant liquor." Carter and other planters "leaned" the bottles so that the cider grazed the corks and kept them swollen. If a cellar had a window, then the planter needed to shut and cover it, usually with straw and dung, to keep the cider "in a proper and temperate state."[46]

The most effective way to preserve cider, however, was to distill it into apple brandy. In the first half of the eighteenth century, those large Chesapeake planters or their wives who wanted to distill cider relied on either drop distilling or rise distilling. In drop distilling, an upper container held the raw materials, and a lower container linked to it collected the alcoholic vapors that boiled off and then cooled, dripping down the pipe into the lower container. Rise distillation was drop distillation reversed. The raw materials were placed in the bottom container, the fire was lit, and the alcoholic vapor was collected in an upper container. Both processes were dangerous, undependable, took hours, and created a liquor that was filled with impurities.

Although distilling had a long history in England, it was still a novelty in the Chesapeake in the eighteenth century. Monks had distilled some alcoholic and nonalcoholic liquors in Britain beginning in the fifteenth century, but the quantities were limited and the process was a secret. In the latter half of the sixteenth century, English doctors began to recommend distilled spirits as everyday drinks. Small numbers of men began to distill commercially. A petition to Parliament in 1621 reported that there were around two hundred alcohol distillers in London. In 1638, English men founded the Company of Distillers, a guild that was not eager to share its techniques.[47]

Chesapeake colonists who wanted to learn more about distilling had to import the books that detailed, or purported to detail, all the distilling innovations. Most colonists could not afford these books, which often cost the steep price of £1. In addition, the books could be difficult to obtain. Before Virginians got a bookstore in the 1760s, planters had to draw on English merchant connections to ship them books on credit and wait ten weeks to one year to receive them.[48]

Only large planters could afford to buy specialized distilling books. A typical small-planter household owned a Bible and an almanac. In contrast, large planter John Mercer owned at least twenty-three books about distilling, cidering, raising hops, or brewing, including William Ellis's *The London and Country Brewer* and Richard Bradley's *Hop Garden*. Landon Carter and Robert Wormeley Carter of Sabine Hall owned *The Experimental Husbandman and Gardener* and *The Practical Husbandman and Planter*, both of which detailed alcoholic beverage production. William Byrd of Westover had George Smith's *Complete Distillery*, John

Worlidge's *Vinetum Britannicum: or, a Treatise of Cider*, William Hughes's *Flower Garden & Vineyard*, and Moses Cook's *The Manner of Raising Fruit Trees* among his books. Many large planters owned John Evelyn's *Sylva, or a Discourse of Forest Trees*, which taught how to make cider and propagate orchard fruit. Chesapeake merchants advertised that they sold *A New and Easy Method of Making Cider Royal*, *The Cider Maker's Instructor*, *Choice Observations Concerning Cider*, and *The Practical Farmer*, which taught, among other topics, how to "make cider keep." Expensive as the husbandry books were, they were common in large planter libraries.[49]

Large planters also subscribed to expensive journals that discussed alcoholic beverage production. William Byrd and Thomas Jefferson owned and read copies of the *Philosophical Transactions of the Royal Society of London*. The Royal Society publications explained new alcoholic beverage production techniques. For example, Richard Reed published his recommendations "for the advancing of cider in richness both for taste and color" in the *Philosophical Transactions*. Planters also found Reverend Henry Miles's instructions "On Some Improvements which may be made in Cider and Perry" and Paul Dudley's "An Account of a new sort of Molasses, made of Apples," which "serves for food and brewing," in the Royal Society's journal.[50]

Stills were extremely expensive. In 1719 a used and possibly broken still owned by a smith who had likely made it himself was valued at £1.5. In contrast, a 1731 probate record listed the value of seven cider casks and a cider tub at only 13s 3d, or less than half the value of the old still. The more typical still cost much more. For example, lawyer and sheriff James Nimmo's "syder still" was appraised at £23 in 1754. Landon Carter imported at least one still that cost almost £40 sterling. Such expenses explain why from 1710 to 1754 only 6.3 percent of households in seven Chesapeake counties had stills.[51]

Families wealthy enough to own stills passed them carefully to their children. Gabriel Alloway bequeathed his son a still as well as at least three plantations. Thomas Samford inherited a still as well as six leather chairs, an expensive desk, and a "great glass," all signs of wealth. William Walker left not only a plantation and a still but also numerous expensive tools. In 1754 James Nimmo bequeathed his still to his son, William, who subsequently passed the still onto his son, James, in 1791. And James Bray III inherited his still from his father.[52]

Importing a still was an aggravating process. It was a smart Mr. Pope who had the still-head and worm he ordered from England engraved with his initials for identification purposes, or "mark'd WP with a B under the P," because he did not know when or on what ship it would arrive. When Young Moreland wanted a brandy still, he had to write to a merchant with English connections for one. Moreland tried to be specific, asking for "a brandy still to hold 60 gallons." But planters received whatever merchants dumped on them. Peter Lyons returned the still that was sent to him "as it leaked, and was so badly tined, that it did not answer my purpose; and desiring you would send me a pewter one." Lyons wrote in frustration that the merchant should "either change your tradesmen or insist on their furnishing you with better goods then they have lately sent your correspondents in Virginia." Despite the difficulties of importing stills, large planters managed to obtain them. Charles Carter of Cleve owned eight copper stills by 1766. His father, Landon Carter, owned at least nine ninety-gallon stills.[53]

Distilling was only for those with a tolerance for risk. The author of an English guide to distilling published in 1718 pointed out that distilling "frequently terminates in the blowing up of the head, or throwing the wash foul in to the worm." Stills frequently broke because of their multiple parts, size, and awkwardness. Small planters complained of "still-burnt" liquor and would not buy it. Philadelphia merchant Israel Pemberton reminded Walter Rossington of this problem, writing that "it is a general complaint" of his distilled liquors that "though it may be strong it is ill tasted & often still burnt which is a great detriment in the sale of it." Some experimenters recommended placing "an iron plate at the bottom of the still with about two inches of good mortar between the still and the plate" in order to "keep the liquor pure and neat" and to "preserve the bottom of the still which would otherwise soon burn through."[54]

Expensive stills tempted thieves as well. Landon Carter was furious to discover in 1767 that parts of his "large new copper still that might work one hundred and twenty gallons" had been stolen. Although Carter offered a £20 reward "hoping the people of this country will use some diligence to detect such a species of villainy," it is not clear that he ever recovered his still parts. Distilling took repeated experimentation with the ability to sustain loss.[55]

Despite the costs and risks, large planters engaged in distilling because of the money they could expect to make. Distilled liquors had the special advantage of being resistant to spoilage. Joseph Ball instructed his nephew, who was managing one of his plantations, to "let the cider be beaten. . . . The rest I would have you get stilled as well as you can, right good and strong." The distilled apple brandy could reach 66 proof (33 percent alcohol), which allowed it to keep. While planters charged between one-half and one shilling per gallon of cider, they charged at least five shillings a gallon for distilled apple or peach brandy. James Bray III, for example, could have charged around £22 for the 86 gallons of distilled brandy he made in 1743 and 1744.

A. *The Still*	L. *A Pewter Crane*
B. *The Worm-tub*	M. *A Pewter Valencia*
C. *The Pump*	N. *Hippocrateis bag or Flannel*
D. *Water-tub*	*Slieve*
E. *A Press*	O. *Poker Fire-shovel Cole-rake*
F.F.F. *Tubs to hold the goods*	P. *A Box of Bungs*
G.G.G.G. *Canns of different size*	Q. *The Worm within the Worm-tub*
H. *A Wood Funnel with a iron-nosel*	*mark'd with prick'd lines*
I. *A large Vessel to put the Fains*	R. *A Piece of Wood to keep down*
or after-runnings	*the Head of the Still to*
K. *Tin-pump*	*prevent flying of*

Large-planter households could also use their stills to make sugar. William Cabell had at least one of his female indentured servants, Mary Duncan, use the sixteen-gallon still that he had purchased from England in 1741 to make sugar from sap at his Warminster plantation in Virginia.[56]

Virginians' consumption of alcohol indicates just how limited were their connections with Atlantic markets. In many ways, large planters were pushed into making distilled beverages and surplus ciders because they could not rely on imports. It was a lucrative business, though, for those who could afford the start-up costs, which included the costs of the English books detailing what to do, the necessary labor and cooperage, and the imported bottles and corks as well as the investment of time.[57]

This analysis of household alcohol production has thus far revealed that women supplied small-planter households with cider during apple season and that during the rest of the year, small-planter households, servants, slaves, and artisans depended on large planters' surplus alcohol output, a production that was overseen by men. Colonists' connection to the Atlantic world depended on their economic class. Most colonists did not engage much with the Atlantic world, and most purchased alcohol produced near home. Large planters, on the other hand, established connections with English merchants and friends who shipped them goods, sometimes including alcohol. The insignificance of imports reveals that Chesapeake colonists were surprisingly self-sufficient as a *community* in producing alcohol. Even when colonists drank at taverns, the subject of the next chapter, they did so in an institution that was far more local than global and was itself beholden to the large planters.

Facing page. Seventeenth- and early-eighteenth-century stills were extraordinarily expensive, large, complex, and fragile systems that were difficult to import. Virginia planter Landon Carter noted that it took "five or six of my best hands" to carry a still (*Virginia Gazette*, 12 March 1767). George Smith, *A Compleat Body of Distilling* (London, 1749), image reproduced with permission of Special Collections, John D. Rockefeller, Jr. Library, the Colonial Williamsburg Foundation.

CHAPTER FOUR

"Anne Howard . . . Will Take in Gentlemen"

White Middling Women and the Tavernkeeping Trade in Colonial Virginia

Taverns mattered in the colonial Chesapeake. Taverns connected colonists spread apart by sprawling tobacco farms. It was in the taverns that colonists learned current crop prices, purchased goods, read newspapers, and discovered business opportunities. Significantly, throughout the seventeenth and eighteenth centuries many, if not most of these, were run by middling-sort women and prominent widows. That so important an institution was frequently managed by women has been unnoticed until now because scholars have agreed that women found their options and roles progressively constricted during the eighteenth century. In fact, women were the preferred tavernkeepers in the eighteenth-century Chesapeake.

Magistrates, who often passed their own positions from father to son, frequently transferred taverns from mother to daughter or to middling families with previous female tavernkeeping experience. Tavern *licenses* were assigned to men, but both magistrates and license applicants knew that the tavern itself would be run by the petitioner's wife or daughter. As a result, most taverns were provincial affairs serving plantation-produced surplus liquor, some rum imported by the planter-merchants, and local food. Passing tavernkeeping through middling women was the magistrates' effort at policing the taverns, not because they were particularly threatened by immorality or other challenges to the social order in the taverns, but because they feared the taverns' potential for rendering colonists penniless.

Magistrates may have hoped that middling women would be a virtuous influence on male patrons. While colonists in the seventeenth and early eighteenth centuries considered women the less moral sex, closer to animals and more easily tempted by the devil than were men, ideas of women's virtue began to change during the mid-eighteenth century. By the latter part of the century, many colonists were assuming that women were the more moral sex, more tender and delicate — the link between humans and angels. Increasingly, women were seen as the ones responsible for inculcating children with self-discipline, benevolence, and piety. The terse court records of the time never state why middling women were preferred as tavernkeepers, but the possibility of their civilizing influence might have affected tavern licensing decisions. After all, as an essayist in the *Virginia Gazette* proclaimed in 1773, women "will ripen the seeds of virtue in men."[1]

Additionally, magistrates awarded taverns to families with financial and social stakes in the status quo in the hope that these tavernkeepers would encourage fiscal responsibility in their patrons. Market forces assisted magistrates in this effort. It was expensive to run a tavern, and the chance of profits was dicey. These factors limited the trade to families who had experience and could bear the financial risk.

Chesapeake magistrates' activities to limit tavernkeeping not only to women but also specifically to middling and upper-sort women with prior tavernkeeping experience may have been unique. In other colonies those obtaining licenses belonged to a great variety of classes. For example, in seventeenth-century Massachusetts, Puritans awarded licenses only to well-off men, believing that the male gentry alone could control the potential violence of taverns. When Anglicans ousted the Puritans from control of the colony after 1720, they also changed who received tavern licenses. After 1720, the new authorities returned to older English ways of awarding licenses to a great variety of people and issued increasing numbers of licenses to poor women in order to prevent the colony from having to support the impoverished. In Philadelphia, Quakers began with many of the same concerns about taverns as Puritans in Boston had but ultimately bowed to the populace's wishes. In 1704 Quaker authorities created a partial, less-expensive license that allowed holders to sell only rum and beer. The licenses were popular among the poorer sort. As a result, in Pennsylvania taverns were run by a wide variety of men and women.[2]

It is important that so many of the Chesapeake's taverns were run by women. In the Chesapeake, taverns were vital, serving the common good in numerous ways. They were spaces where men shared news and sold goods and slaves, where strangers and visitors rested, and where people gathered to discuss politics and crop prices and to retrieve their mail. A tavern could even serve as a lost-and-found. In 1768 Anthony Hay advertised that he had "lost on the evening of the 8th instant, somewhere nigh Mr. Robert Nicholson's in this city, a small silver watch," and he offered a reward for the watch's return to his tavern. Groups of men shared subscriptions to the newspapers that they read, often aloud, in the taverns. As a sign of this practice, one Virginia taproom posted a notice stating that "gentlemen learning to spell are requested to use last week's newsletter." Letters were deposited at taverns for colonists to collect. Daniel Fisher, a Carter plantation tutor, commented that "a letter directed to John Palmer esq. at Williamsburg" left at Chiswell's tavern, "lay upon a table, which several persons who were going thither viewed, but neither of them took the trouble of conveying it as directed; a common neglect, it seems, unless the person has a mind to see the inside of the letter, a practice often complained on." Taverns were sites of information dispersal in other ways as well. When the catastrophic yellow fever epidemic hit Philadelphia in 1794, news of it spread to Annapolis when "Gab[rie]l Duvall and W[illia]m Pinkeney both came" to Williams Farris's tavern to "say that there is a fever raging there [Philadelphia] that carried 40 or 50 a day."[3]

Taverns were also critical as social venues in a place where colonists lacked the church ales, gentry festivals, and other sponsored pleasantries of England and New England. Taverns occasionally connected the southern colonies to European high culture. Jane Vobe once exhibited paintings by Mr. Pratt of England in her King's Arms tavern in Williamsburg. Anne Shields arranged an English-style ball at her Williamsburg tavern in 1751. Williamsburg's Raleigh tavern offered some classes in the "art of fencing, dancing, and the French tongue" in 1752. Colonists played games in the taverns. Anne Shield's tavern had "four packs of cards"; Jane Vobe's tavern contained "six dozen best hary cards"; and Mary Hunter's tavern offered ninepins, cards, and a gaming table. Catharine Jennings kept "an exceeding good billiard-table" in her Annapolis tavern. Taverns provided spaces for more active entertainments as well. Horse races often began and ended at taverns, as did militia training exercises. Cockfights took place in the taverns, like the "grand cock fight" of a New Castle, Virginia, tavern. Taverns provided the occasional exotic excitement as well. For instance, the Borough Tavern in Norfolk, Virginia, advertised in 1793 that "a beautiful African lion can be seen at their tavern."[4]

Gentlemen formed social clubs that met in the taverns. For example, Robert Wormeley Carter recorded that he paid five pence for club-related tavern expenses. The investors of the Pennsylvania Vine Company, a group viniculture experiment, met to discuss company business at the Rotterdam tavern. Dr. Alexander Hamilton's Tuesday Club, a drinking group, met frequently in taverns. Phi Beta Kappa, the literary and debating society founded in 1776 at the College of William and Mary, met in Williamsburg's Raleigh Tavern.[5]

Taverns, or "ordinaries," served so many social functions that an anonymous clergyman complained in the *Virginia Gazette* in 1751:

It is notorious, that ordinaries are now, in a great measure, perverted from their original intention and proper use; viz. the reception, accommodation, and refreshment of the weary and benighted traveler; which ends they least serve or answer and are become the common receptacle and rendezvous . . . where prohibited and unlawful games, sports, and pastimes are used, followed, and practiced, almost with any intermission; namely cards, dice, horse-racing, and cockfighting, together with vices and enormities of every other kind.[6]

Taverns served political purposes as well. George Washington used Christiana Campbell's tavern for political club meetings from 1762 to 1774 when debating the Townsend Duties, the Stamp Act, and other imperial policies. Thomas Jefferson did the same at Jane Vobe's tavern in 1768 and 1769 when he met with lawyers, merchants, ship captains, and local men to discuss the news from abroad. Mary Davis advertised her offer to keep "a table for 10 or 12 Burgesses" so that the debates concerning political and legal decisions might take place in her tavern.[7]

Finally, taverns played a critical economic role. Given the lack of towns, markets, and shops, many economic transactions transpired in the taverns. A study of Elizabeth River (Norfolk), Virginia, found that in 1648 at least 80 of the town's 334 men owed money to one of the area's tavernkeepers. In addition to borrowing money from the tavernkeeper, colonists bought and sold goods in the taverns themselves. For example, James Geddy, a Williamsburg silversmith, sold his jewelry at Bennett White's tavern. Tavernkeeper Jane Vobe offered "riding chairs, both double and single, with harness new and complete, a small tumbrel, two carts, nine very good cart horses, with harness, [and] several men's saddles and bridles" for sale or rent at her Williamsburg tavern. Colonists occasionally purchased alcoholic beverages in taverns to take home. For example, when William Byrd ran short of wine during a surprise visit from friends, he bought a bottle at a nearby tavern for their visit. Likewise, when Mrs. Geddy needed wine for her husband's funeral, she bought some at Anne Pattison's tavern in Williamsburg. Brissot de Warville, a traveler visiting the colonies, concluded that "the innkeeper himself is a respected man in a country where money is scarce, for more of it passes through his hands than anyone else's."[8]

Operating a tavern carried social and economic responsibility, so tavernkeepers were selected carefully. In the eighteenth-century Chesapeake, only middling- or upper-sort families ran legal taverns. A close examination of tavern licenses reveals that only well-connected middling families whose women had previous tavernkeeping experience or a claim to the trade received tavern licenses. First, families that received licenses were well-off. A survey of Virginia taverns indicates that magistrates limited licenses to the middling sort (artisans, small planters, prominent widows, and government employees) and the well-connected. Colonel Armistead and Colonel Beverley, both wealthy planter-merchants, erected taverns.

A French traveler to the Chesapeake noted that a tavern he stayed at "is kept by one of the most respectable families in Maryland." The widow of wealthy Virginia planter Edmund Pendleton kept a ferry and a tavern. Lewis Burwell, a wealthy and powerful planter who held the prestigious and lucrative position of naval officer for the Upper James, had a ferry, a ferry house, a warehouse, and a tavern, all on his Kingsmill plantation during the 1730s and 1740s. Daniel Fisher obtained "a [tavern] license at the county court whereof he is himself a member" in 1767. Prominent widows applied for licenses successfully. The widow of John Luke, a planter who had been collector of customs, kept a popular tavern in Williamsburg in 1710. Likewise, the widow of Commissary William Dawson, who had been president of the College of William and Mary, owned a tavern.[9]

Indeed, in some ways taverns were extensions of Virginia's large plantations. Taverns that were not owned by magistrates and planter-merchants outright still depended on them for operating licenses, alcoholic drinks, and customers. Barriers to entry made tavernkeeping the domain of the planters-merchants and those they favored. The cost of the tavern license, the requirement in some counties that a wealthy man pledge security for the proposed tavern owner, the magistrates' preference for awarding licenses to men married to women with previous tavern-keeping experience, the expense of keeping a tavern, and the uncertainty of profits, kept tavernkeeping in the hands of relatively few families.

Eighteenth-century records reveal one of the magistrates' unrecorded rules of petitioning successfully for a tavern license: the petitioner had to be backed by a large planter or planter-merchant if possible. (In frontier areas, there may not have been large planters or planter-merchants to support petitions.) Some counties required that someone on the court or known and respected by the court pledge security for the proposed tavernkeeper. The security pledged to cover the tavernkeeper's debts, so he had to trust the proposed tavernkeeper, and the courts had to have confidence in them both. This tactic successfully discouraged the lower sort from applying for tavern licenses, which may have been the point. A survey of orders, wills, and deeds of twenty colonial Virginia counties has uncovered no records of a man who pledged security having to cover a tavernkeeper's debts, suggesting that county courts who required security did so to discourage unacceptable applications.

Examples of security from Surry County, Virginia, indicate how well-connected Virginia tavernkeepers were. Most counties did not record the names of those who pledged security; fortunately, Surry County did. For instance, Edmund Ruffin successfully petitioned for a tavern license in Surry at least eight times from 1742 to 1750. It is not surprising that he was repeatedly awarded tavern licenses.

The Ruffins had been in Surry since at least 1685, when Robert Ruffin received a 2,250-acre grant in Lawne's Creek Parish, Surry. In addition, Edmund Ruffin's mother, Elizabeth, owned 3,001 acres as of 1704, making her the fourth-largest landowner in the county. His father, Robert Ruffin, held the lucrative and prestigious position of sheriff of Surry by 1714, shortly before his death. Ruffin's mother's death quickly followed, and his brother, John Ruffin, inherited the bulk of their parents' estate, with the rest being divided equally among the remaining children: Joseph, Benjamin, Edmund, Mary, Martha, and Elizabeth. Edmund, therefore, came from wealth but needed to make a living.[10]

Needing to make a living was not enough for the court. More important to the justices were Edmund Ruffin's connections. His brother John, a member of the House of Burgesses, was the first to swear security for Edmund. In the following years, Edmund Ruffin also presented William Edwards, Augustine Claiborne, and Benjamin Briggs as his securities.[11]

Edmund Ruffin's securities were from families more prestigious and powerful than his own. William Edwards's family had been in Surry County since at least 1657, when William Edwards (I) was granted 490 acres by patent. Edwards served as clerk of Surry County, a profitable position, and his holdings had swelled to 2,290 acres at the time of his death. His son, William (II) inherited the land, who willed it to his son William (III). William Edwards III was granted at least eight additional patents in Surry County, amounting to 5,120 acres, before he died in 1698. William Edwards IV inherited nearly 4,000 acres, which he gave to his son, William Edwards V. It was William V (1714–71) who stood security for Edmund Ruffin. The Edwards' estate, Pleasant Point, was relatively close to the Ruffins, with a 180-acre property separating the estates on the Ruffin's west side. In a region where twenty acres was the minimum for a viable farm, 180 acres, about one-third of a square mile, was a relatively close distance between neighbors.

William Edwards V enjoyed more than land and wealth; he was a trusted man in the colony, serving as a justice of the peace and as a tobacco inspector. In 1743 William Edwards was made under-sheriff. The sheriff and under-sheriff positions were desirable, and application for them was competitive since men in these positions could charge for delivery of each arrest, peace bond, subpoena, summons, and whipping. The court trusted Edwards' assessment when he vouched for his neighbor, Edmund Ruffin.[12]

Another of Ruffin's securities, Augustine Claiborne, was an eminent man as well. Augustine Claiborne's father, William Claiborne Sr., was the largest landowner in Sussex County, next to Surry, and owned between 3,000 and 4,000 acres along both banks of the Nottoway River. Augustine Claiborne inherited much of this land. When he married Mary Herbert, the only heir to her father's 15,000-acre estate, he grew even wealthier, and his estate, Windsor, finally measured at least several thousand acres. With wealth came power, and Claiborne was a burgess for Surry County from 1748 to 1754 as well as the Surry County clerk from 1751 to 1754. Claiborne clearly approved of Ruffin, acting as his security from 1745 to 1748 and again from 1749 to 1751. And with Claiborne's approval came that of the court; the Surry justices approved Edmund Ruffin for a tavern license.

In another example, Howell Edmonds applied for and received a tavern license from the Surry County court for his charge, John Edmonds, in 1744. John's father had died, and although his mother was alive, the courts required that "orphans" (children without fathers) have a court-appointed male guardian until the age of twenty-one. Like the Ruffins, the Edmonds family was well-established in Surry. Howell Edmonds had married into the Blunt family, and in 1701 his father-in-law, Thomas Blunt, conveyed Shingleton plantation to Howell Edmonds. Howell added to this when he was awarded a 1,004-acre patent on the Blackwater River. At the same time, the Surry County courthouse moved to Shingleton plantation, leaving the Edmonds conveniently situated to have a tavern "at the courthouse."

The Blunt family, into which Howell Edmonds had married, was wealthy and well-regarded in Surry county, with Thomas Blunt (Edmonds' father-in-law) serving as both sheriff and captain of the militia. Colonel Richard Blunt was a member of the House of Burgesses from 1772 to 1774. Not surprisingly, given the prestige of the Edmondses and the Blunts, when Richard Blunt offered to act as security for John Edmonds' tavern, the court approved.[13]

The case of John Edmonds' tavern license reveals an unwritten and unrecognized requirement for obtaining a tavern license: that the practice of tavernkeeping remain in certain families, passed on by women. Both Edmonds and Ruffin had a tavernkeeper in the family: John Edmonds' mother, Ann Edmonds, had operated a tavern with her first husband, Thomas Edmonds. When Thomas Edmonds died, Ann Edmonds married Edmund Ruffin. It was only after this marriage that Ruffin applied for and received a tavern license. It is no coincidence that Ann Edmonds' husband and son both received licenses. Although it was often men who applied for and received the licenses, tavernkeeping was a family business frequently performed by women because of their experience and because the men ostensibly "running" the taverns typically were busy with other jobs. Though most licenses were in male names, the tavern itself was frequently operated by wives and daughters.

Regardless of whose name was on the license, it was often, if not usually, a woman who managed the tavern, as advertisements make clear. Anne Shepherd

Images such as this one have obscured women's roles in managing taverns
from view. The partially hidden woman in the background carrying
the turkey is likely the tavern's owner or manager. *The Country Club* by
Henry Bunbury, engraved by William Dickinson, London, 26 June 1788.
Image published by permission of the Colonial Williamsburg Foundation.

placed a notice after her husband died in 1748 that "I [will] still continue to keep
my house up the path, for the entertainment of such gentlemen and ladies as are
pleased to favor me with their company." Mary Davis advertised in 1773 that she
had "several large and small rooms for the entertainment of ladies and gentlemen." Grissel Hay publicized her offer of "very commodious lodgings to let for a
dozen gentlemen, and their servants, with stables and provisions for their horses"
in Williamsburg. Anne Pattison announced that, "as she proposes keeping a very
genteel plentiful house, she hopes to meet with encouragement." Christiana
Campbell published the announcement that her tavern was moving to "the
house, behind the capitol, lately occupied by Mrs. Vobe; where those gentlemen
who please to favor me with their custom may depend upon genteel accommodations, and the very best entertainment."[14]

Whether a tavern license was in a male or female name, it was typically
women who managed the taverns. Wills, tax lists, and advertisements show that
male "tavernkeepers" were busy managing other businesses. In Williamsburg,
wigmaker Richard Charton; cabinetmaker Anthony Hay; silversmith John Coke;

tailor Thomas Craig; harnessmaker Gabriel Maupin; gaoler Jean Marot; cabinet-maker Matthew Moody; surveyor, juryman, and tobacco inspector Vincent Rust; and lawyer John Markland all owned tavern licenses. Their wives handled the day-to-day management of the taverns. For instance, when Matthew Moody's wife died, Martha Drewitt applied for his wife's former business even though the license was in Matthew Moody's name. Women often inherited the management, if not also the licenses, of their mothers' taverns. For example, when Elizabeth Marriott died in 1755, her daughter, Anne (Marriott) Howard, assumed her mother's tavern. Howard advertised that "Anne Howard, (living at the sign of the ship where her mother formerly kept tavern, in Annapolis, and) having a number of very good spare beds and bedding, and a convenient house for entertainment, will take in gentlemen." Christiana (Burdett) Campbell, whose tavern in Williamsburg is now a popular tourist attraction, grew up the daughter of a tavernkeeper. When her husband died, she returned to her hometown of Williamsburg to reopen the family business. Mary Maupin, the wife of Gabriel Maupin, the previously mentioned harnessmaker who received a tavern license from the court, had kept a tavern before she married.[15]

Beyond handing down taverns through the female side of a family, women were critical to the process of securing a tavern license from the court. For example, Anne (Marot) Sullivant initially ran a tavern with her first husband, Jean Marot, the town gaoler, in the early years of the eighteenth century. When Jean Marot died, Anne continued to operate the tavern. She later married Timothy Sullivant, whom the court approved for a tavern license, and he helped to maintain Anne Marot's tavern while continuing his work as a stonecutter. When Anne Marot died, she passed the tavern on to her daughter, Edith Marot, who ran the tavern with her husband, Samuel Cobbs. Likewise, Mary Maupin and Gabriel Maupin managed a tavern. When Gabriel died, Mary married Thomas Crease, and they opened a tavern together. Thomas, who had no previous tavernkeeping experience and was busy with his position as gardener for the Governor's Palace and for the College of William and Mary, obtained a license because of Mary's experience. Similarly, Hadley Ravencroft received a license after marrying Dionisia Ravencroft in 1705. He retained his position of superintendent of buildings for the College of William and Mary, and she managed their tavern. Tavernkeeping women sometimes engaged in multiple professions themselves, but their second jobs were more conducive to operating taverns. For example, Jane Vobe ran a tavern in Williamsburg and sold foodstuffs to the Virginia militia during the American Revolution. Other tavernkeeping women took in sewing, did laundering, or served as the occasional midwife.[16]

"Men's" taverns were thus frequently managed by wives and daughters. Jean Francois Marquis de Chastelluex noted in the journal he kept during his travels to America that "breakfast [at the tavern] was served by Captain Paxton's daughters." John Bramham acknowledged his wife's management of their tavern when, in his 1754 will, he requested his "bro[ther] Benjamin to take the lease of my ordinary on his own account discharging my wife from any trouble concerning it." Planter William Byrd wrote of his stay at a tavern that the tavernkeeper, "Mrs. Newton . . . apper'd to be one of the fine ladys of the town." Francisco de Miranda, touring the country, appreciated the efforts of "Comfort and Constance," the daughters of a tavernkeeper who cooked, cleaned, and perhaps provided other services at a tavern. Miranda recorded, "That evening there was a good supper and better conversation with the girls; after I had retired for the night, one had no embarrassment in coming at my request to continue the conversation in my bed." When one traveler to Virginia's frontier was forced to stay in a male-run tavern, he complained grievously about "this castle of poverty and male accommodations." When men were unavailable or uninterested in maintaining licenses, for example, during the wars that were frequent during the eighteenth century, the courts awarded licenses to women. And when the husband of a tavernkeeping couple passed away, the courts usually agreed to grant his license to his widow. When Elizabeth Leighton's husband died, the Yorktown court awarded her a tavern license in her own name.[17]

Many men who inherited or owned taverns had married tavernkeeping women. Henry Wetherburn, a well-known Williamsburg tavernkeeper, twice married tavernkeeping women. He first married tavernkeeper Mary Bowcok. Ten days after Mary Bowcok died, Wetherburn married Anne Shields, who had grown up helping her mother, Anne Marot, run a tavern and later operated a tavern with her first husband, James Shields. Men often stipulated in their wills that their womenfolk or female friends who were experienced with tavernkeeping should inherit and operate their taverns. When David Cunningham, a tavernkeeper from 1713 to circa 1717, died, he requested that since his wife had passed away, his children and his tavern should be maintained by his "loving friend," tavernkeeper Susanna Allen. Since David Cunningham was a barber and a constable, Susanna Allen most likely had been running the tavern since the death of Cunningham's wife.[18]

Tavernkeeping passed through women's hands in other ways as well. John Timberlake was able to operate a tavern before he died in 1714 because he rented the property from tavernkeeper Susanna Allen. Christopher Ayscough was permitted to purchase a tavern license from 1768 to 1770 because of his wife's skills. Anne Ayscough was a highly respected cook for Governor Fauquier, and when she left Fauquier's service, he bequeathed her £250 for her fidelity and attention. This gift allowed the Ayscoughs to purchase the tavern and the necessary goods, and Anne Ayscough's service for the governor gained her a reputation that Christopher Ayscough used to promote their tavern in Williamsburg. He advertised in the *Virginia Gazette* that "Mrs. Ayscough very well understands the cookery part." The success of the Ayscoughs' tavern, frequented by Thomas Jefferson, was probably owed more to Anne Ayscough than to Christopher. While she ran the tavern, he was serving as a doorkeeper and tipstaff to the Council. The Council ultimately dismissed Christopher Ayscough for repeated drunkenness. Christopher was allowed to enter the tavern trade because he had married Anne. She provided the capital, the skills, and the steadiness to operate the tavern.[19]

This analysis raises the questions of both why and how Virginia's magistrates limited tavernkeeping licenses. To begin with the *why*, Virginia's magistrates and legislators wanted sober, self-supporting, and productive colonists. English tradition dictated that parish residents who were too poor or indisposed to support themselves had to be provided for with local taxes. If indigent numbers increased, so did taxes. From the laws legislators passed, it appears they feared that, left to their own devices, many colonists would drink away their livelihoods and increase the poor rolls. Without a police force, magistrates' only hope to control colonists' activities in the taverns was to select tavernkeepers who would uphold the upper sorts' interests.

Virginia laws concerning taverns and tavernkeeping focused on their potential for financial irresponsibility. Only infrequently did they pass laws regarding morality. Despite regular presentations of colonists to the courts for "tippling on the Sabbath" in both the seventeenth and eighteenth centuries, legislators gave up on trying to prevent such drinking by law after the 1690s. The issue was not raised again until 1748, when legislators requested only that tavernkeepers prohibit drinking and gambling on the Sabbath "any more than is necessary." Legislators passed no other laws related to issues of taverns and morality in the eighteenth century.

Legislators concentrated on the potential of taverns for economic disruption to the colony. They attempted to limit those whom tavernkeepers could permit to drink in the taverns. For example, in 1691 the legislators declared that it was against the law to sell to "any person or persons, who are not master of two servants, or being visibly worth £50 sterling at least." Legislators apparently feared that the lower sort would harm the economic interests of their masters or drink away their resources in the taverns. In 1705 legislators similarly declared that tavernkeepers were not allowed to sell to seamen because their wages instead "should be support of themselves and families." In 1734 the legislators declared that tavernkeepers were culpable for the lower sorts' poverty, since it was the "evil practice of ordinary keepers, and retailers of wine, beer, rum and other distilled spirits, and mix'd liquors, selling such liquors upon credit, to the impoverishment and ruin of many poor families." Legislators warned tavernkeepers not to expect the courts to help them recover debts from these sales. In short, legislators feared that tavernkeepers would allow the lower sort to drink away their livelihoods and add to the poor rolls or hurt the productivity of their masters.[20]

The legislators were not against the sale of liquor to the upper sort. Nor were they against the sale of alcohol so long as merchants, and not the tavernkeepers, stood to profit. The legislators made clear that the colony's merchants, often including magistrates and legislators themselves, could sue in court to recover debts. It was only tavernkeepers who were not allowed to recover debt from sales. Moreover, when the court set limits on how much tavernkeepers could sell, they made exceptions to favor the upper sort. For instance, the legislators declared in 1734 that sales limits did not apply to tavernkeepers in Williamsburg during the quarterly general court meetings. Legislators, magistrates, planters, and merchants flocked to Williamsburg during the general court meetings and wanted to be able to drink. The legislators' intention was to prevent the lower sort from spending in the taverns, not to limit their own imbibing.[21]

Frequently, though, tavernkeepers did not follow the law. Legislators often had to repeat and rewrite laws because tavernkeepers were not abiding by them. For example, the court repeated the laws prohibiting tavernkeepers from selling hard liquor in 1644, 1646, and 1647 because the tavernkeepers sold liquor despite the law. In 1658 the court implicitly acknowledged that its attempts to limit such sales were unsuccessful and agreed that tavernkeepers could sell liquor, but only at the high prices set by the court in order to restrict sales. If a tavernkeeper sold alcohol for more or for less than the rate the court ordered, then the tavernkeeper would be "fined double the value of such rates so exacted."[22]

Tavernkeepers resisted laws that limited their clientele and the quantities they could sell until the legislators passed a new law in 1762. This one stated that tavernkeepers could sell liquor "to any person whatsoever, except sailors in actual pay on board any ship, or such person as shall be actually inhabitants of the county," in any quantity the tavernkeeper thought fit. The purpose of this law was again to prevent local citizenry from landing on the poor rolls. The court said that tavernkeepers could sell as much as they wanted to strangers and travelers but curbed sales to locals whose families the upper sort might have to support if drinking sapped their savings.[23]

Over the years, courts required tavernkeepers to agree to various rules. For example, they might be required to post court-approved rates for food and drink and face fines if they deviated from these amounts. However, justices did not enforce the fines. Even when tavernkeepers allowed egregious behavior, the courts were reluctant to interfere. For example, although Susanna Allen, from 1710 to 1719, operated a Williamsburg tavern known among residents as "bawdy" and the site of at least one riot, she was not punished by the courts.[24]

In practice, colonial authorities did not enforce the tavern laws. Perhaps the public outcry would have been too great. Instead, magistrates focused on limiting tavern licenses and hoped for the best. Given that the justices pinned their hopes on tavernkeepers, it is startling that a review of twenty counties uncovered no cases in which the court denied a license. It is possible that denials were not recorded. It is more likely, though, that the high cost of the license and of keeping a tavern limited tavernkeeping to those whom the court found acceptable. In the one case of an illegal tavern recorded by the surveyed courts, the court forgave the family for selling alcohol after they promised not to sell without a license again, a result perhaps based on the family's poverty. The justices then asked the governor to consider the family "an object of charity."[25]

Still, it is not possible to ascribe the low incidence of unlicensed tavernkeepers to colonial laws alone. Market forces were equally responsible for this vetting, which, again, scholars have not noted. The cost of the tavern license and of the trade itself prohibited most colonists from entering the market. By 1705 a proposed tavernkeeper had to "petition the county court" and give a bond of ten thousand pounds of tobacco for a license. Selling alcohol without a license became punishable that year by the forfeit of two thousand pounds of tobacco or the acceptance of twenty-one lashes at the public whipping post. In 1748 the court made such licenses good for one year only and required that the courts consider "the convenience of the place proposed, and the ability of the petitioner to keep good and sufficient houses, lodging, and entertainment for travelers, their servants and horses" before awarding a tavern license. By 1748 licenses required a bond of £50 current money to the King and 35s current money for the use of the Governor. Or, as visitor Peter Kalm concluded, "no one is allowed to keep a public house without the Governor's leave; which is only to be obtained by the payment of a certain fee." The £50 35s annual fee kept a tavernkeeping license out of the reach of many colonists who earned on average £11 a year.[26]

More expensive than the license was outfitting the tavern. Most tavern records from early Virginia have been lost, but a 1752 to 1758 ledger from the Ship Tavern, probably owned by Charles Carter of Cleve in Falmouth, Stafford County, Virginia, has survived. Charles Carter (1707–1764) was the wealthy son of Robert

Carter of Carter's Grove, a man who had for years purchased land along the Potomac and Rappahannock rivers for his sons, Charles and John. When Robert
Carter died in 1732, Charles Carter inherited 13,000 acres in Stafford County and
decided to move there from Urbanna. Sometime after 1737, he bought a plantation known as "Cleve" from Ralph Wormeley and had built a brick mansion on
the land by 1750, thus giving him the name "Charles Carter of Cleve." Wealthy
and headstrong, Charles Carter of Cleve owned at least one hundred slaves and
served as both a justice and a county lieutenant for King George County. In August of 1736 he was made a burgess for King George County, a position he held
for the next twenty-eight years. By 1752 he had opened the Ship Tavern in Falmouth.[27]

Running the Ship Tavern was not cheap. The Ship Tavern ledger begins in
1752, and its early records say nothing about the start-up costs of the tavern or
whether Carter had to build the tavern. Carter did, however, make a partial list in
1754 of his tavern inventory and costs. Items on the list indicate that the tavern was
new in 1754, since Carter purchased basic cooking supplies and beds. The list included:

> 7 feather beds, Negro Moll, jack, cart, 11 pairs sheets and pillow cases, 8 pair
> blankets, 2 colored rugs, 4 spotted, 1 Torrington rug, 2 dozen pewter plates and
> dishes, 2 pair tongs, 1 shovel, tablecloths and towels, 1 tea kettle, 1 frying pan, 1
> colander, a spade, 2 gill, 1 half pint pots, a looking glass, knives and forks, 3 wal-
> nut tables, iron pots, 1 pewter chamberpot, 1/2 dozen chairs, 1 sifter, pepper box,
> 1 mustard pot, 2 salts, 1 vinegar cruet, 1 curry comb and brush, 2 brass corks, 1
> bow and yearling, 1 tin coffeepot, 1 horse.

These goods cost Carter at least £160.5.10, according to his records. In addition
to "Negro Moll," Carter employed at least one other servant for the tavern not included in the list, Jacob Eteston, to whom he paid £50. The cost of the tavern, not
including the building, alcohol, food, or other supply costs was thus at least £260.

Moreover, Carter had to be patient for profits. The year 1755 offers the best period for analysis because its records are the most thorough. By the end of 1755, customers had incurred debts of £515 at the Ship Tavern. The fact that customers
owed Carter this amount does not necessarily mean that he received it. As of 1761,
when the account book ends, seventy-one customers, or 40 percent had yet to pay
their debts, although six years had passed. The number of customers who had not
paid may be higher than that, since cases where the customer "says that he has
paid," or had made a partial payment by 1761, were counted for this analysis as
having paid. The length of time it took for men to pay such debts limited who

could go into tavernkeeping. Lower-sort families did not have enough wealth to extend so much credit to their customers.[28]

Of the customers who paid their debts, some paid in cash, but many paid in goods and services, which meant that a tavernkeeper had to know how to dispose of a variety of goods profitably. Carter received payments in bootstraps, shoes, corn (twenty barrels of it one day), lambs, chickens, wheat, fodder, butter, eggs, and turnips. Some of these items Carter could use in the tavern, particularly for stews; the rest he probably traded for other items he needed. In addition, a few customers spent several days working for Carter in order to pay off their bills. Of Carter's customers, 49 percent paid with cash or notes of credit they had received from other colonists, almost 11 percent paid in goods or services, and, again, 40 percent simply had not paid six years after accumulating their debts and perhaps never paid at all.

Before the profitability of the Ship Tavern can be ascertained, it is necessary to know more about Carter's expenses. The Ship Tavern ledger does not list the expenses of food and drink for customers, fodder for their horses, or the other products and services that Carter provided. James Barrett Southall's 1771–1775 tavern ledger, probably from Williamsburg's Raleigh Tavern, indicates what expenses were like in a more cosmopolitan area. The items from the ledger shown in table 1 cost the tavern owner £344 over the five years that the accounts were kept. If this number were applied to Charles Carter, then supplies would have cost him £69 per year. As a result, for the year of 1755, Charles Carter would have made only £30 if his non-paying customers did not meet their bills. Unless he was able to collect from his non-paying customers, Carter may have earned less from tavernkeeping than his servant did.[29]

Charles Carter of Cleve was not the only tavernkeeper whose customers left outstanding debts for years. Such a situation was typical, and part of what kept

TABLE 1

Raleigh Tavern Food and Stable Expenses, 1771–1775

oats	£ 6	beef	£ 6 2s 4p
lemons	£ 5	pork	£ 27 6s 9p
fodder	£ 3	coal	£ 10 8s 4p
bread	£ 42 54s 4p	cheese	£ 10 8s 4p
butter	£ 32 12s 6p	wine	£ 70 43s 6p
arrack punch	£ 8 8s 9p	beer	£ 45 12s 16p
sugar & limes	£ 7 2s	porter	£ 36 44s
gin	£ 7 10s	coffee	£ 3 2s 6p
rum	£ 16 13s		

James Barrett Southall, "Accounts 1768–1771," photostat, John D. Rockefeller, Jr. Library, Williamsburg, Virginia.

tavernkeeping the domain of the well-off was the frequent necessity of threaten-
ing an action to recover debts. For example, tavernkeeper Elizabeth Leighton-
house did not collect on three debts from 1701 until she took the cases to court in
1707. Tavernkeeper Susanna Allen took Timothy Metcalf to court for a long-
outstanding debt of £12. Mary Frazier announced in the *Maryland Gazette* in August of 1748 that she "desires all persons indebted to her to discharge their accounts immediately, or else to give notes for the same, which will prevent trouble
to themselves." Evidently, her customers did not pay their debts, because six
months later Frazier pronounced again that she "hereby desires all persons indebted to her, to come and pay off their respective debt, by the end of this month,
or they may expect to be sued." Jane Vobe also threatened legal action if her delin-
quent tavern customers did not pay. Ann Tilly advertised that people who purchased liquor and provisions at her tavern during the general election in December of 1767 needed to pay their debts, "as the gentlemen who stood as candidates
at that election, have refused to pay me." John Ward took John Bolling to court
for a debt that had not been repaid in ten years. Customers stayed at taverns knowing that they might be unable to pay. When Lucious Bierce toured the South with
a friend, he knew in York, Virginia, that "we had not money to pay our board and
did not like to own it." So he told the tavernkeeper that "we would prefer having
our accounts stand till the close of the first quarter," hoping that they would find
a way to pay their bill when due.[30]

If recovering debts took such persistence, why did men and women continue
to enter into tavernkeeping? The number of people who requested tavern licenses
indicates that tavernkeeping either was, or was perceived to be, profitable over the
long term. Moreover, tavernkeeping brought other benefits, like the acquisition
of market information and some political power. For example, after Edmund Ruffin received a tavernkeeping license from the court, he was appointed and paid
by the court to appraise several estates, to act as a guardian for two orphans, to
build a local prison for which he was paid £60, and to serve as the county gaoler,
for which he received 508 pounds of tobacco.[31]

Moreover, many taverns were family endeavors. Although men obtained many
of the licenses and may have fronted the fees, it was often women who managed
the taverns. Operating taverns allowed women to contribute to family incomes
and to socialize with neighbors and visitors regularly. Women could run taverns
out their homes while they raised children, oversaw servants and slaves, grew herbs
and vegetables, dipped candles, sewed clothes, washed laundry, made cider, and
cooked dinner.

For widows, tavernkeeping offered a livelihood. It was almost impossible for a woman to run a farm by herself. She could hire help, but there was no guarantee that hired men would act in her interest. Tavernkeeping offered women a chance to move into town and sociability. For example, after Christiana Campbell's husband died, she moved to Williamsburg and maintained a tavern there for seventeen years. Jane Vobe operated the King's Arms tavern in Williamsburg for more than thirty years. Tavernkeeping provided Campbell and Vobe with a respectable way to make a living.[32]

Perhaps it was because colonial Chesapeake taverns were run by middling women who were usually unable to travel, or even to read, that they were generally such provincial, local affairs. In the Chesapeake, taverns were usually run out of the proprietor's home. In contrast, taverns in Europe, New England, and the Middle Colonies were more often free-standing businesses. The taverns in Williamsburg, the city that became the initial United States capital, offered more in the way of English entertainments and European furnishings and drinks than was typical for the Chesapeake. But even the Raleigh Tavern, perhaps the most wealthy and cosmopolitan of the Chesapeake's taverns, was mocked by European visitors. "There is the Raleigh tavern," sneered Scotsman Alexander Macaulay shortly after his arrival in Yorktown, "where more business has been transacted than on the Exchange of London or Amsterdam; in that building formerly assembled the rich, wealthy merchants of all countries from Indies to the pole, from the Tweed to the Orades; here the exchange of the world, the relative value of money in every kingdom on earth was settled." Most taverns would have met with far more censure from Macaulay. Landon Carter noted that a tavern in Westmoreland county had "not a pane to the windows, and the very ballroom was presently covered with snow." "Great want of beds," Nicholas Cresswell noted at Lynch's Tavern, "but I tent with the floor and my blanket."[33]

Tavernkeepers sold mostly either alcoholic drinks and other foodstuffs produced on the plantations of Virginia's large planters or foodstuffs they made themselves, again keeping taverns unsophisticated. Susannah Allen purchased cider for her tavern from William Byrd's plantation. Carter's Grove plantation supplied Anne Pattison's tavern with cider and wood. She bought more wood and wine from the Littletown plantation. Christina Campbell bought beef, veal, and wheat for her tavern from William Lightfoot's estate. And John Powell outfitted his tavern with butter, meal, and cider from the Granbery plantation. The homemade food and drink led more cosmopolitan travelers to complain. "Breakfasted at Rollin's, a public house," Nicholas Cresswell began his observations on taverns, "in this country called ordinaries, and indeed they have not their name for nothing, for they are ordinary enough. Have had either bacon or chickens every meal since I came into the country. If I continue in this way shall be grown over with bristles or feathers."[34]

Anne Pattison kept a well-to-do tavern in fashionable Williamsburg. Like other tavernkeepers, she served food and drink that she and her servants had made on the premises or wine and rum she had purchased from large planters. She brewed and served her own molasses "beer" as well as butter and cheese she made herself. She served vegetables from the tavern garden, hominy beans, and stews made of the chickens and beef with which customers paid her. To be sure, most taverns served wine, rum, and rum punch made with ingredients imported by large planters or planter-merchants from the Caribbean. These items were popular. And fashionable Williamsburg taverns had red-checked curtains and tea services, as tavernkeepers tried to keep up with fashions in England. But the majority of accommodations, food, and drink, at most Chesapeake taverns was decidedly local.[35]

Throughout the eighteenth century, taverns served critical social and economic functions. Many of these taverns were run by women of middling families who had the support of the region's large planters. Indeed, much of what was sold in the taverns came from large-planters' farms. Small-planter households, tired of depending on taverns and large planters for year-round drink, though, were about to find an explosion of places to purchase alcoholic beverages.

CHAPTER FIVE

"Ladys Here All Go to Market to Supply Their Pantry"

Alcohol for Sale, 1760 to 1776

After the middle of the eighteenth century, Chesapeake colonists found an increasing variety of places where they could buy alcoholic beverages. They were no longer limited to producing alcohol at home, buying it from the surplus of large planters, or purchasing it from taverns favored by large planters. Especially after 1760, colonists could purchase all sorts of alcoholic drinks at markets, Scottish factor stores, or other shops in small towns. Among the new choices was one alcoholic drink that had long been inaccessible to most Chesapeake colonists: rum. As colonists flocked to the new places to buy alcoholic beverages, large planters desperately endeavored to retain their old command over small planters and laborers by expanding their operations and experimenting with new kinds of drink. Because of declining crop prices and financial crises during the latter half of the eighteenth century, large planters needed the income that selling alcohol provided more than ever. However, almost all of their efforts failed. Colonists preferred to shop in the new venues and enjoyed their increasing independence from large planters.

It was not until the latter half of the eighteenth century that Chesapeake colonists began to build marketplaces where they could purchase goods such as alcoholic beverages. In the eighteenth-century Chesapeake, marketplaces were large open-air buildings divided into rental stalls, providing space where goods were guaranteed to be for sale. Although marketplaces were highly useful, their development was delayed by Chesapeake colonists' focus on the tobacco crop. Tobacco was so labor-intensive, and separated colonists so widely, that colonists refrained from building markets or marketing networks. County governing bodies tried to promote town development, but until the middle of the eighteenth century, their efforts were unsuccessful. For example, when Governor Argall returned to Virginia in May of 1617, he found that the land that he had set aside for a marketplace in Jamestown had been planted with tobacco. In 1649, the privilege of creating a weekly market was offered to Jamestown (then James City), but the townspeople lacked interest and the marketplace went unbuilt. In 1699, a student of the College of William and Mary asked Governor Nicholson and the Assembly to establish a marketplace in Williamsburg for the ease of the students and the College, but again there was little interest. In 1705, the Assembly voted to create weekly markets in Williamsburg without results. In 1736, Norfolk merchant John Taylor spent £46 to build a market house, but he could not find anyone to rent space from him.[1]

Chesapeake colonists finally began to establish marketplaces in the 1750s. Colonists in Fairfax County, Virginia, raised money to build a marketplace in 1751. Annapolis colonists took bids for their marketplace in 1752. In 1757, plans were finally implemented in Williamsburg for a market house. Norfolk expanded its marketplace in 1764, and Baltimore rented out a marketplace for the first time in 1765. Portsmouth established a marketplace in 1783, Petersburg in 1784, and Yorktown in 1786. Once built, the marketplaces were popular. Most held one or two market days per week. By the late eighteenth century, some marketplaces, such as Norfolk's, were open every day. Colonists found pork, lamb, mutton, poultry, livestock, eggs, milk, butter, cheese, and alcoholic drinks available for purchase. Fishermen sold fish and shellfish. Bread and grains were often available as well. By 1770, as Mary Ambler commented in Baltimore, "ladys here all go to market to supply their pantry."[2]

Williamsburg established an especially large marketplace, and the activities that took place reduced the dependence of small planters on large planters. In 1769, regional merchants advertised a plan to hold four merchant meetings a year in Williamsburg. Ship captains, insurers, agents, and customers all descended upon Williamsburg during the quarterly meetings to buy and sell. The routinization of the merchant meetings made it easier for small planters to order goods from European agents directly, begin to purchase shipping insurance, set crop prices, and trade with others. As Governor Fauquier wrote, the meetings brought "persons engaged in business of any kind" to the capital in order "to expedite the mode and shorten the expense of doing business."[3]

Much as the building of marketplaces and the formalization of the merchant meetings in Williamsburg gave small planters autonomy, so too did the legislatures' passage of the 1730 Tobacco Inspection Act, however unintentionally. The Tobacco Act aimed to preserve the quality and esteem of Virginia tobacco by requiring that all tobacco intended for export at public warehouses be inspected. The act required the construction of tobacco warehouses to hold tobacco for inspection every twelve to fourteen miles on Virginia's navigable rivers. If a farmer's tobacco passed inspection, then he was given a tobacco note stating how much the tobacco was worth. Colonists could use the notes to pay taxes, and tobacco notes soon became a standard medium of exchange. The Tobacco Act thus created a currency that encouraged trade between people of all social standings, reducing the dependence of small planters and laborers on loans and book credit from large planters. A second consequence of the Tobacco Act was that the buildings constructed to house the tobacco became sites of general trade.[4]

The most important result of the establishment of the tobacco warehouses along the rivers of the rural Chesapeake during the 1760s was the advent of the Scottish stores. Scotland had been little involved in the tobacco trade before 1707, because England had prohibited direct trade between Scotland and the English colonies. When Scotland united with England in 1707, Scottish merchants began to assume a large role in the tobacco trade. Scottish merchants had several advantages. Glasgow was two to three weeks' sailing time closer to the Chesapeake Bay than was London since Scottish ships could sail north of Ireland rather than south of it. This route also offered Scottish merchants safety in times of war. The Scandinavian countries and France, all of which enjoyed strong trading ties with Scotland, were increasing their demand for the Oronoco tobacco grown in the Chesapeake, and they turned to Scottish merchants for it.[5]

The first Scottish merchants in the Chesapeake, traveling merchants sent over to hawk goods, were often called "Scots peddlers." In the mid-eighteenth century however, Scottish tobacco firms began sending over permanent representatives called "factors," who would insure that their employers would get enough tobacco. To persuade colonists not to sell to the large planters, the Scottish factors offered both goods for trade and long-term credit to buy those goods. Scottish factors opened stores near the tobacco inspection warehouses where colonists could purchase goods on credit and pay for them later in the year with their tobacco crops.[6]

The Scottish stores were especially popular in rural Chesapeake areas, where they competed successfully against large planters. Many Scottish firms opened numerous stores, lowering prices. For example, James & Robert Donald & Co. of Greenock had six stores in Virginia by 1763. By 1774, the Cuninghame firm had seven stores in Maryland and fourteen in Virginia. One of the benefits of the early "chain" store system was that when one store ran out of an item, the factor could restock it from an associate store. Having multiple stores allowed the factors to buy in bulk and to cut the amount of time that their ships idled in the Tidewater. The head factor would distribute goods from the ships himself instead of relying, as the planter-merchants did, on the ships' captains to distribute the goods. This quicker turn-around meant that the Scottish factors received two annual shipments from Europe rather than the planter-merchants' one shipment. As more factors set up year-round stores, the increasing competition led factors to increase the credit they offered on tobacco in order to woo small planters' business. The factors came to depend on the sale of goods for their profits.[7]

The Scottish stores had a reputation for friendliness that the planter-merchants did not. Merchant James Robinson urged his new Scottish storekeeper to give "all good usage and drink in abundance." The effect, as one traveler wrote, was that farmers "carry their tobacco to their merchants for which they barter for such commodities or goods as they want, they who live 100 miles and upwards in the back country, they lay in goods [that] will serve them until the following year." The new Scottish stores opened up a world of goods for the Chesapeake's small planter households, who elected to patronize the new stores for their variety of goods, favorable terms of credit, low prices, and friendly atmosphere, to the detriment of the large planters.[8]

One of the new goods in the Scottish stores was rum. The popularity of rum in the stores both reflected and encouraged a social trend that further reduced colonists' dependence on large planters, especially after 1760, when rum and its ingredients became cheaper. While rum had been well-liked in the Chesapeake since the colonies' foundings, it was not until the latter half of the eighteenth century that most colonists could afford the drink. In 1710, rum cost eight shillings per gallon in Virginia, while Virginia cider cost just under one shilling per gallon. By 1752, the price of rum had dropped to five shillings per gallon, and by 1774 it was just four shillings per gallon. Rum provided more alcohol than cider by volume, and since rum was generally mixed with water, a gallon of rum provided more drinks than a gallon of cider.[9]

Another advantage of rum was that it did not spoil in the hot Chesapeake climate as wheat and fruit brews did. Daniel Roberdeau, who founded a distillery in Alexandria, Virginia, in 1767, wrote to a friend that "the most sober inhabitants plead the absolute necessity of rum in the excessive heat of our summer when even cider they say will not be sufficient for men in the fields." Sales of rum had an inverse relationship to the availability of peach and apple brandy, and Virginia merchant William Allason warned his investors that rum would not sell when planters had bumper crops of peaches and apples.[10]

Factor John Hook's store in the backcountry of New London, Virginia, provides an example of the new method of shopping at factor stores and the popularity of rum. Hook sold casks of brandy and whiskey as well as related items such as glass goblets and decanters. His best-selling product was rum, which sold daily between 1760 and 1810. Rum totaled 6.5 percent of the store's total sales. Customers purchased rum by the quart in one and one-half shilling installments. If they did not have cash or tobacco credits, they sometimes paid with a day's work in Hook's garden, two raccoon skins, a yard of homemade cloth, or eight chickens.[11]

Alcohol comprised a greater proportion of factor store purchases for the lower sort than it did for the upper sort. Five percent of the lower sort's total purchases by value were alcoholic drinks, mostly rum. Hook's wealthy customers, on the other hand, devoted only one percent of their expenditures to alcoholic drinks, none of which was rum. Wealthy planters had slaves make their alcoholic drinks, imported alcohol directly from Europe or the West Indies, and looked upon rum as a drink of the laboring classes.[12]

Other businesses and shops began to develop near the Scottish factor stores, making it ever easier for small-planter households and artisans to buy and sell goods on their own. By the later eighteenth century, Virginians could find shops at Cabin Point on the James River, Urbanna on the Rappahannock River, and Dumfries on the Potomac River. By the late eighteenth and early nineteenth centuries, small towns had developed around the tobacco warehouses and shipping sites, offering multiple shops where goods and foodstuffs were for sale: Richmond, Falmouth, Fredericksburg, Alexandria, and Norfolk are examples. In Alexandria, Virginia, for example, the *Alexandria Gazette* advertised regular auctions where apple brandy, French brandy, rum, gin, whiskey, and wine were for sale. William Ransay on Prince Street sold wine, port, claret, rum, brandy, gin, and cognac. On King Street, James Bacon had a "grocery, tea, wine and liquor store" with wine, claret, rum, brandy, gin, and whiskey available. William Hartshorne sold wine; Ricketts & Newton sold Jamaica spirits; Thomas Simms advertised gin, brandy, Jamaica spirits, wine, and barley; J & T Vowell sold quarter casks of port wine from arriving ships. A. Willis offered cider by the barrel and the bottle. James Sanderson had "for sale genuine old port wine in pipes and bottles," as well as Jamaica rum, Holland gin, and brandy. The front page of the August 4, 1803, edition of the *Alexandria Gazette* listed eight merchants selling alcoholic beverages. The number of venues where Chesapeake colonists could buy alcoholic beverages had exploded.[13]

Small producers of alcoholic beverages proliferated in the latter eighteenth century. When tobacco prices dropped in the 1760s, planters began experimenting with raising wheat. The availability of wheat allowed some men to attempt to make ale and beer, particularly as hops became more available in the region in the mid-eighteenth century when New England colonists began growing and ex-

porting hops to the southern colonies. In the absence of refrigeration, commercial attempts to sell ale and hopped beer were short-lived; wheat brews spoiled
more quickly and were trickier to produce than fruit ciders. Nevertheless, by the
latter decades of the eighteenth century, Chesapeake colonists had multiple
choices about where to buy their alcoholic beverages.

Small-planter households and small businessmen were enormously interested
in producing and selling home brew. Newspaper advertisements in the *Virginia
Gazette* show some of their activities. John Mayo advertised that, "intending to set
up a brewery and distillery at this place [Manchester, Virginia], I have provided
two stills and three coppers, one of which contains about 300 gallons." Mr. Bradley of Petersburg, Virginia, "wanted immediately," a man familiar with "the malting and brewing business." George Goosley, near Yorktown, promised good wages
for "a man acquainted with malting and brewing," in 1778. Joseph Kidd, a recent
immigrant, announced that he had "just arrived from England, capable of making malt and conducting a brewery, [and] would be glad of encouragement in the
above branches from any gentleman, or company of gentlemen. He is ready to
pledge his all, that he could bring the above branches to as great perfection as they
are in Britain." Likewise, Richard Smaddell sought "an overseer's place, or a
brewer's place, thinking himself capable of either." Two commercial distilleries
sprang up in Norfolk, two in Alexandria, one in Annapolis, one in Charles Town
(then part of Virginia), one in Baltimore, and one in New Castle County, Delaware, just over the Maryland line. In addition, hundreds of farmers opened small
operations, like Philip Watson, who distilled from two stills on his Williamsburg
farm and sold part of the resulting alcohol to his neighbors.[14]

While individual production was small, Chesapeake colonists, in the aggre-
gate, were distilling extraordinary amounts of fruit alcohols and rum. Ten percent
of the rum that they produced was exported, which left an annual average of
twenty-one gallons of Chesapeake-produced rum for every adult white male. By
1770, the average white male drank more than three pints of rum per week, or
nineteen and one-half gallons per year, or the equivalent of seven one-ounce
shots of rum every day of the year. The popularity of rum in the American colonies is highlighted by a comparison with that of England and Wales, where in
1770 average per capita consumption of all distilled spirits was only six-tenths of a
gallon. The Chesapeake was drenched in alcohol.[15]

Large planters, at least those whose records remain, were unhappy to lose the
income and stature associated with providing alcohol to small planters, artisans,
and laborers. Faced with the defection of their patrons, some large planters des-

perately increased their production of alcoholic beverages for sale on their plantations. They hoped to make a profit from alcoholic beverages in a declining economy and to keep their small-planter customers.

Chesapeake men interested in alcoholic beverage production for economic diversification found their hopes for profit echoed in alcoholic beverage production manuals in the later eighteenth century. Earlier instructors had offered only the vaguest suggestions of profits. For instance, one author noted only that after "eight or nine years of growth of the trees, prodigious profits have thereby accrued to their economy." Most early writers did not mention profits at all, instead emphasizing the opportunity for quality control. For example, the author of the seventeenth-century *The Complete English Brewer* offered "easy directions for brewing all sorts of malt liquors in the greatest perfection" so that more discerning drinkers would not have to drink alcohol made by "the usual method, where the person is not nice or delicate in his malt-liquor." Later-eighteenth-century writers gave stronger assurances of riches, however, and began referring to alcoholic beverage production as a "business." Increasingly, authors stressed that "there is no business in which a small capital will yield so large a profit."[16]

Large and small planters constantly feared that tobacco prices would drop or that the weather would damage the crop. "It is more uncertain for a planter to get money by consigned tobacco," lamented William Fitzhugh, "than to get a prize in the lottery." All too frequently, planters found themselves warning that "no great crops will be this year made, by reason of our great drought." Robert Beverly apologized to Landon Carter for not repaying several large sums of money because, "the unfavorable season, we have had for some years, will not enable the most industrious or opulent planters in the country to comply so punctually with their engagements as they might wish." John Baylor concluded that selling his tobacco was "hardly worth negotiations," and that tobacco would soon "fall too low for us poor planters to live by." Richard Corbin unhappily concurred. "The sale of my tobacco last year," he wrote, "f[e]ll very short of my expectations," and he warned soberly that "the crops next year will be still shorter." Another time John Custis predicted that "we had such a violent flood of rain and prodigious gust of wind, that the like I do believe never apprehended since the universal deluge it has destroyed most if not all the tobacco in the colony."[17]

Although planters tentatively expanded into other crops such as wheat in the second half of the eighteenth century, this did not offset their fears. Francis Taylor found that growing wheat was as unreliable as growing tobacco. "Wheat crops are generally sorry—weather very dry," he noted. "Half the wheat," recorded

George Washington, "(and some say a great deal more) and 3/4th of the rye, are blasted." Later, too much rain destroyed Washington's corn as well. John Baylor concluded that the "sale of indigo are so far below my expenses" that it was not worth his trouble to grow it. And the same storm that destroyed John Custis's tobacco also left his corn "fodder most ruined."[18]

Concerns about the weather and tobacco markets had long beset planters, but the situation during the 1760s became dire. During the late 1750s drought had destroyed most crops, and planters became increasingly indebted to the European merchants from whom they purchased their goods. Moreover, soil exhaustion had reduced tobacco output per laborer. Despite the declining crops and profits, however, after mid-century planters began spending *more* on imported goods as they indulged their desires for consumer goods. The planters' situation began to reach emergency proportions during the 1760s, especially when international financial crises during the 1760s and 1770s compelled British merchants to call in long-standing debts.[19]

Faced with economic troubles, Chesapeake colonists could not simply diversify their holdings in the eighteenth century. The only means of communicating business information was by letters sent on lengthy and precarious voyages. Planters and merchants went for months with little to no information on the markets. The resulting information asymmetry made business contracting difficult. Additionally, capital markets were ill-developed until the early national period, so planters could not purchase securities of various firms to protect their economic stability. During the 1760s, leading colonists argued that manufacturing and progressive agriculture were the panacea for an ailing economy, and the home-manufacturing movement gained momentum. Particularly after the passage of the Townshend duties in 1767, which placed taxes on imports such as lead, paper, paint, glass, and tea, home agriculture and agricultural experimentation, including alcoholic beverage production, became fashionable for men in the colonies.

Large and small planters were forced to look for ways that they could reduce their expenditures or supplement their incomes in the second half of the eighteenth century. Chesapeake planters, both large and small, spent tremendous amounts of money on alcohol. Gentry families generally allocated 20 percent of their food budgets to alcohol; they spent more only on meat. For example, in 1782 Thomas Jefferson spent $80 for beer, $50 for cider, $30 for whiskey, and $225 for wine (equivalent to at least $9,050 in 2007), in addition to costs for brandy, porter, and other alcoholic drinks. Smaller planters often purchased alcoholic beverages that they could not make themselves from merchants and factor stores. Virginia merchant Francis Jerdone on the Paumonky River estimated that he could sell at least 4,000 gallons of rum. But falling tobacco prices meant that after the mid-eighteenth century planters could ill afford these costs.[20]

Hopes of profits from alcoholic beverage production were not limited to sales of alcoholic drink alone. By-products of alcoholic beverage production, particularly the spent-grain wash, could be fed to cattle and hogs. John Fitzgerald, who brewed, reminded George Washington of the "amazing benefit your stock of cattle and hogs will receive" from the used wash. Brewers and distillers placed their mash tubs near their cattle and hogs in order to garner the profits of fattened livestock. John Taylor, author of *Arator*, the most influential agricultural work in the late-eighteenth-century colonies, warned potential ciderers that "distilling from fruit is precarious, troublesome, trifling and out of his province," but he reminded planters that "the apple will furnish some food for his hogs, a luxury for his family in winter, and a healthy liquor for himself and his laborers all the year. Independent of any surplus of cider he may spare, it is an object of solid profit."[21]

Increasing slave holdings and slave reproduction made it easier for Chesapeake planters to engage in alcoholic beverage production. During the first half of the eighteenth century, nearly all slaves were agricultural laborers. Even Robert "King" Carter, who owned 390 enslaved persons of working age in 1732, had most of his slaves in the fields. But as planters increased their slave holdings, and as the enslaved community reproduced itself, planters began to place some slaves in craft positions in order to create work for extra hands, reduce dependence on British imports, and have the convenience of goods made on the plantations. Until the 1730s there were not enough slave women in the Chesapeake for slaves to reproduce themselves, so planters had to use nearly all of their spare capital to buy more slaves. Once slaves were reproducing, planters no longer had to direct so much of their capital toward this end. Moreover, as American-born slaves, who spoke English, replaced African imports, it was easier for planters to begin to train slaves in craft skills. By the 1770s two-fifths of the male slaves in the Tidewater had some artisan training. Planters searching for ways to use their increasing numbers of slaves fully, and with some freed-up capital after mid-century, had the labor and money to put into alcoholic beverage production. For example, George Mason's father taught some slaves to distill in the 1770s. George Washington had at least four slaves trained in distilling. And Thomas Jefferson hired help to train some of slaves to brew.[22]

Living with increasing debt and needing to use their slaves fully, planters explored the potential of plantation alcohol production for profit. Some men had modest goals for profits to be realized from selling alcoholic beverages once the region had enough population in the second half of the century to support selling liquor. For instance, in 1751 William Dudley in Richmond County hoped only with the "money to be raised out of the drink which is made on this plantation any time within five years" to build a small house for his son. Other planters had larger dreams. A few examples illustrate the spectrum of alcoholic beverage production activities in which planters engaged.[23]

John Mercer of Marlborough attempted the first known large Virginia plantation brewery in 1767. Mercer was one of the colony's most prominent and wealthy attorneys. His oratorical skills earned him almost £300 as an attorney in 1730; in 1731 he earned almost £700. John Mercer earned an extraordinary amount of money for an eighteenth-century colonist; yet, like many planters, he lived beyond his means. In the 1740s Mercer began spending lavishly, buying fine clothes and expensive jewelry, having his portrait painted, and replacing his four-wheeled chaise, first with a carriage and then with a chariot. The new chariot led to new horses. In the late 1740s Mercer began building Marlborough, his plantation manor house. Marlborough boasted a Palladian facade, stone corner quoins, and brass hardware. Mercer filled Marlborough with mirrored sconce glasses, marble-topped sideboards, and imported furniture. He paid the Atlantic passage of an English gardener to design the grounds and hired plasterers, woodcarvers, stone carvers, and a master carpenter. John Mercer's Marlborough was an enormously expensive showpiece.[24]

But Mercer did not have the money to pay for the new goods and the mansion, despite his income. In part this lack was due to his difficulty securing payments from his clients. "Since the first day of last January," he fumed in his journal, he had not been able to collect "£10, out of the near £10,000 due me." Mostly, though, Mercer simply spent more than he earned, especially as profits from tobacco crops declined. When increasing deafness forced Mercer to retire from the bench in 1765, he needed to make his plantation profitable.[25]

In 1765 Mercer decided to commence large-scale plantation brewing for profit and hired William Baily to be his brewery overseer. In addition to Baily, Mercer also hired "one Wales, a young Scotch brewer." He wrote at length to his son that he had decided to build a brewery that "would quickly retrieve all my losses and misfortunes." Otherwise, he feared that he would be "reduced to the most necessitous circumstances." Mercer hoped that his brewery would produce enough beer to sell to ordinaries, meaning taverns. He estimated that "our ordinaries abound and daily increase (for drinking will continue longer than anything but eating)," and he dreamed of profit. Mercer hoped that at least brewing his own beer would allow him to provide "brew for my family [including slaves] use, that they may have drink with their victuals," even if he could not provide for them in other ways.[26]

John Mercer had developed an astonishing vision for his brewery by this point. He built a brewhouse and a malthouse out of brick and stone, each 100 feet long, and a cooper's house. He wrote to his son that he had built a new hand mill in the brewhouse loft that would grind fifty bushels of malt each morning, and that he was hunting for an expert malter. Mercer planned to grow his own barley and hops and to hire his own permanent coopers to build casks for shipping his beer. He was determined to add a glass-bottle manufactory to his plantation so that he could bottle his beer for sale. And he wanted to ship the beer himself with his own vessels, made in his own shipyard, sailing from his own dock. These were not idle daydreams. He wrote to his son with the terms on which he would accept a shipping partner, asked his son to search for such a man, and explained how he would rent out the ships when they were not wanted for transporting beer. He advertised to the public that he "propose[s] setting up a glasshouse for making bottles, and to provide proper vessels . . . at the several landings they desire." He purchased an additional forty slaves to grow the barley, set up the cooperage, and malt and brew for him, which increased his investment in his brewery to over £8,000. His neighbors watched his activities and wrote to Thomas and Martha Jefferson that Mercer was transforming into "a complete farmer . . . who will scarce believe it."[27]

Mercer's dreams never came to fruition. Bailey's beer was so bad that it could not be sold, a typical outcome for breweries in the South until the invention of refrigeration, and Wales brewed so little that Mercer could not recoup the expenses of his salary and maintenance. Though Mercer continued to believe that he could construct a profitable brewery until the day he died, his brewery failed utterly. The only people he managed to convince to buy his beer were people who owed him money. John Mercer's debt, and that of his spendthrift oldest son George, passed onto Mercer's son James, who was burdened with the debt for the rest of his life. In 1768 Mercer's wife, Ann, and their sons began selling off the estate to pay the debts. Still, Mercer's unsuccessful brewery illustrates the expanding array of alcohol varieties and venues in the region, as well as the part that financial desires and enslaved labor played in large planters' escalation of alcoholic beverage production.[28]

George Washington's alcoholic beverage production efforts fared little better. Washington, too, was motivated by economic desires. Washington suspected as early as 1775 that the duties he undertook for the new United States during the American Revolution would come at a personal financial loss. At the start of the war he instructed Lund Washington, his overseer at Mount Vernon, to observe "the greatest economy and frugality; as I suppose you know that I do not get a far-thing for my services here more than my expenses; it becomes necessary, there-fore, for me to be saving at home." Washington was correct: his years of service were not financially rewarding, and he retired to Mount Vernon deeply worried about his financial situation.[29]

Washington knew that his neighbor, John Fitzgerald, occasionally distilled molasses into rum, and he wrote to Fitzgerald to ask whether a large plantation distillery might be profitable. The answer evidently led Washington to conclude that building a distillery "will rather increase my labor," but that labor "will never be any objection where a saving can be made." He hoped that his distillery, "under the uncertainty of cropping of late years, would, with good management and close attention to them, be found my best and most certain support." In July of 1797, Washington hired a Scottish farm manger named James Anderson to start a plantation distillery. "Mr. Anderson has engaged me in a distillery, on a small scale, and is very desirous of increasing," Washington wrote to a neighbor, "assur-ing me from his own experience in this country, and in Europe that I shall find my account in it . . . the thing is new to me."[30]

Like John Mercer with his brewery, Washington soon invested considerable amounts into his distillery, adding three more stills, another boiler, $520 in addi-tional mash tubs, building a still house and a malting house, and planning to add a cellar, until he had a large operation. These expenses forced Washington to call in some debts owed to him. As he explained to Henry Lee, he was sorry to call in Lee's debt, but "my own want requires it," because "I have encountered a consid-erable expense in building a large distillery." Washington complained to his friend Robert Lewis that the distillery "has cost me a considerable sum already, but I find these expenditures are but a small part of the advances I must make." The distillery added to Washington's financial pressures rather than relieving them, and he railed at Anderson for "having induced me to encounter a very se-rious expense in erecting a distillery" until Anderson threatened to quit. Though the two men apparently smoothed over some of their differences and maintained the distillery, the operation was not a success. Washington died in 1799 without seeing his distillery turn a profit. (To be fair, interpreters at Mount Vernon plan-tation say that Washington's distillery did earn money, but they do not account for his expenses in erecting and running the distillery and include sales of pork and beef from animals fed with spent-grain wash as distillery profits.) Washington's dis-tillery again indicates large planters' hopes to make profits from selling a greater variety of alcoholic beverages produced on the large scale.[31]

Virginia planter-merchant Daniel Roberdeau similarly engaged in alcoholic beverage production in hopes of making profits. When Roberdeau was threatened with the bankruptcy of his West Indies rum trade in 1773, he pledged that "like a snail I have retired within my shell . . . to secure a competency" and a "most laudable provision" for his family at his Alexandria, Virginia, plantation. Roberdeau decided to set up a large distillery in order to build a nest egg for his family and an inheritance for his children. By 1774 Roberdeau was asking merchants Mess. King and Harper to send as much molasses as their ship could carry and had engaged £3,000 in building his distillery. Roberdeau hired an overseer named David Jackson to run the distillery, engaged merchant Isaac Winn to supply him with "a constant supply of molasses," and announced that he planned to distill 1,500 hogsheads of molasses per year. Roberdeau wrote that he intended to sell the rum he made from wharves and a "range of stores" that he would build and own. He estimated that the distillery would clear £2,800 profit per year.[32]

The American Revolution disrupted Roberdeau's plans, and he wrote to Thomas Jefferson that he had purchased only "one small still of 500 gallons to help support the necessary expense of my family until the present conflict shall subside." Roberdeau never made a success of his distillery, and by 1781 he had declared bankruptcy. His distillery equipment and materials were sold at a sheriff's sale. Despite its sad conclusion, Roberdeau's story illustrates the enthusiasm for rum and the rise of a variety of producers and locations to buy alcoholic drinks by the late eighteenth century.[33]

Small planters' decisions to purchase alcoholic beverages from the growing array of stores or from the improving apple varieties and presses forever reduced their dependence on large planters for alcoholic beverages after 1760. The fact that this cidering and distilling was increasingly performed by men rather than by women is the subject of the next chapter.

CHAPTER SIX

"Every Man His Own Distiller"

Technology, the American Revolution, and the
Masculinization of Alcohol Production in the
Late Eighteenth Century

The majority of Chesapeake men became interested in the traditionally femi-
nine project of making alcoholic beverages during the second half of the eighteenth century. A very small number of men, all of them the largest planters, had
already become involved with large-scale alcohol production, but they were the
exception in the region. Most men in the Chesapeake lagged behind men in Europe, Latin America, New England, and the Middle Colonies, who had been
making alcoholic beverages since at least the seventeenth century. The transition was not speedy, and rural women in western areas of Virginia and Maryland
continued to produce alcohol well into the nineteenth century. In more eastern
and populated areas, however, the penetration of English science, the advent of
technological advances, the necessity of supplying the Continental Army with
liquor during the American Revolution, and the compliance of women with
men's efforts led to the masculinization of alcohol production in the second half
of the eighteenth century.

Beginning in the late seventeenth century, "scientific" Englishmen throughout the English empire exhorted their brothers to take up the production of alcohol. For example, Dr. Fothergill of the Bath Agricultural Society in England
urged "every gentleman who wishes to improve his estate" "to be well versed, at
least, in the principles of philosophical chemistry," in part because the "brewing,
the making of wine, cider, vinegar, &c. are so many chemical processes; which,
for want of the requisite stock of knowledge, in many cases either fail altogether,
or are carried on with little advantage." English scientists and authors rewrote
women's recipes into precise directives intended for men, stressed the difficulty of
making the beverages, and in general transformed alcohol production from cookery to science.[1]

The new instructors began by insisting that men follow their newly calculated commands. For instance, Eliza Smith's 1727 *The Compleat Housewife*, which was extremely popular in both England and the colonies, never told women how to make malt. Smith assumed that women had this knowledge and that the process was simple enough. In contrast, Alexander Morrice told his male readership in his 1802 *A Treatise on Brewing* that making malt required them to place the barley in water for seventy-two hours; drain it for thirty hours; stir it every three to five hours; heap it for twelve hours, turn it every six hours; and then place the barley in a kiln for four to twelve hours. New experts like Morrice emphasized that brewers should keep "a book of different brewings, and observation thereon." "I must recommend," ran the usual instructions to scientific brewers in the eighteenth century, "that he will never make a brewing without keeping a correct account of his day's work."[2]

Experts added newly coined scientific terms including "alkali" and "narcotic" to alcohol production, published analyses of specific gravities of the products used in making alcohol, and insisted that each batch of brew be considered a scientific "specimen." In harmony with this advice, Virginia planter Landon Carter wrote in his journal in 1772 that "every husbandman whether planter or farmer would do well to keep a diary or journal of all his observations on his own and the management that he sees of others; for as it can never be perfect, it is certain he himself might correct many of his own errors by such a journal comparing one year with another."[3]

The new instructors of the late eighteenth and early nineteenth centuries aligned alcohol production with the emerging field of chemistry. "It is only by working as closely as possible to the principles of chemicals science," proclaimed one author, "that the best and most profitable result can be produced" in making alcohol. Many instructors recommended that brewers and distillers read William Irvine's *Chemical Essays*. The author of *Hall's Distiller* quoted long sections from Mr. Henry's "Epitome of Chemistry" and explained that alcoholic beverage production was "dependent" on chemistry, with "great advantage to be derived from a knowledge of this science." Making alcohol, the new experts insisted, was an activity within the field of chemistry, not cookery.[4]

At the same time, authors of scientific tracts assuaged any fears of complexity their readers might have had by labeling their systems of alcohol production "practical." Alexander Morrice declared that he was a "common brewer" with "practical abilities." The notices published to promote Harrison Hall's book insisted that he was a "practical man" with "practical information" of "practical good sense." Likewise, authors stressed the word "plain": John Tuck assured readers that his book was "on a plain and entire new plan," and promised that he would not "introduce any thing that is not easy to be comprehended by the plainest understanding."[5]

Overall, the treatises celebrated the idea that science had transformed women's art and mystery into men's certainty. For centuries, English men and women had declared that making alcohol was a woman's skill. For instance, cookbook author Robert May stressed in 1660 that preserving and distilling were "secrets" that belonged, as the title of his cookbook proclaimed, to the *Art and Mystery of Cookery*. As late as 1727, some English cookbook authors were still stating that women were "artists in the brewing way" and should use their expertise "to judge as they please." Considered as a science, however, alcohol production was entirely knowable; the mystery was removed. The new science authors emphasized that men should "master" women's "mystery." "Though I shall give every information in my power of the criterion by which to judge when a perfect fermentation has taken place," one scientific author stressed, "nothing but practice and your own observation can make you master of it." "It is absolutely necessary to endeavor to be a master of this knowledge," another writer told his readership, almost all male. The best beer would be brewed by "whoever will make himself master of these lessons," proclaimed a third author. The few authors who persisted in invoking the ancient art and mystery of alcohol production were criticized by their peers. Authors rebuked John Richardson for his "reprehensible [book], on account of the air of mystery in which the subject is invested."[6]

Sometimes authors evoked the art and mystery of alcohol production in order to claim that they could reveal secrets that others could not. This was itself another way of demonstrating mastery. Alexander Morrice promised that his book "exhibited the whole process of the art and mystery of brewing," indicating that brewing was no longer a mystery to *him*. Since he would show "the manner of using the thermometer and saccharometer" "rendered easy to any capacity," he established himself as a master of the mystery. Other authors similarly referred to the artfulness of the brewer to enforce the notion of mastery. Brewing, these authors assured, "requires the strictest attention of the artist." Authors increasingly emphasized alcoholic beverage production "mastery" and science as the process was more and more de-skilled, to discourage women's participation and to assuage men's concerns. A man who had "mastered" alcoholic beverage production did not need to feel ashamed of performing women's work.[7]

In case any men continued to leave alcohol production to women, the new experts assured them that they wrong. Morrice warned that "when a butt wants fining down, [many] appoint a servant girl to perform that office by whom the bungs are left out, and many other acts committed, which all tend to discredit the brewer, although he does not deserve it." In answer to those who pointed out that "every old woman can brew," Morrice argued that women, "not knowing the proper heats that are necessary," "are giving goods instead of grains to the pigs." Women, the new experts asserted, could no longer differentiate between barley and beer.[8]

The transition to all-male, scientific alcohol production in the Chesapeake began with the large planters, who, eager to keep up with the gentry in England, communicated with the Royal Society of London and read its publications. The Royal Society, founded by Francis Bacon in 1660, was the first scientific society in the English empire. Scientific societies such as the Virginia Society for Advancing Useful Knowledge, founded in 1773 "in humble imitation of the Royal Society," used Francis Bacon's writings as a guide. Planters wrote long letters to the Royal Society describing regional flora and fauna, and sent samples of snakes, corn, and seeds. Planter Philip Ludwell, owner of the Green Spring plantation, wrote to the Royal Society in April 1760 to request grape slips for winemaking. Virginia Governor Francis Fauquier corresponded with the Royal Society as well, and in 1762 planter Charles Carter sent the Royal Society samples of his wine. John Clayton of Virginia co-authored with a European a 1739 book on American botany called *Flora Virginica*. John Leeds of Annapolis published his observations of the transit of Venus in the Royal Society publication, *The Philosophical Transactions*.[9]

Large planters saw scientific distilling as a way to keep up with the English gentry. Philip Vickers Fithian recorded a conversation at Mr. Carter's dinner table between men who were speaking about distilling persimmon "beer." "It is soft, mild, of a fair pure color, burns clear, but does not answer the Colonel's expectations," Fithian noted, "so that he does not propose to recommend it to his neighbors in this or the neighboring counties as a useful experiment." Charles Carter's experiments with producing wine suggest the same scientific view. "I am collecting seed from all parts of the country," he wrote to the Royal Society in England, but "our grapes seldom take root, the joints being longer than all foreign grapes that I have seen."[10]

Elite Chesapeake men wrote their own expert literature. For example, by 1775 Virginia planter Landon Carter was experimenting with brewing from green corn, recording his actions with each brew and trying to improve the result each time. Once he thought that he was successful, he published his instructions for green corn beer in the *Virginia Gazette* so that other households could adopt his techniques.[11]

It is difficult to determine when and how Chesapeake colonists below large-planter status absorbed the recommendation that alcohol production become a male science. Only a few Chesapeake journals have survived, and even they contain mostly brief entries on weather and crops. Almost nothing written by women still exists. The few journals and careful readings of other remnants such as cookbooks and husbandry books reveal that Chesapeake men began assuming the task of producing alcohol in increasing numbers during the second half of the eighteenth century.

Recipes and instructions slowly migrated from cookbooks to husbandry books that small planters bought. Cookbooks that Chesapeake women purchased and wrote in the seventeenth and early eighteenth centuries emphasized that alcohol production was women's responsibility. Remember that Sir Kenelme Digbie devoted the initial one-third of his famous 1669 cookbook for women to alcoholic beverage recipes. Richard Bradley marketed his popular *The Country Housewife and Lady's Director* (1727) by advertising in the subtitle that it contained "instruction for managing the brew house, and malt liquors in the cellar; the making of wines of all sorts . . . [and] practical observations concerning distilling" for women. "The reason which induces me to address the following piece to the fair sex," Bradley explained, "is, because the principal matters contained in it are within the liberty of their [women's] province." Bradley included recipes for women to make birch wine (birch syrup, yeast, water, lemon peel, sugar, raisins, and cloves), other raw alcoholic beverages, herbal distillations, and ales. Bradley assured women that he meant no disrespect in giving them alcoholic beverage instructions, stressing that "artists in the brewing way are at liberty to judge as they please." Other cookbook authors popular in the Chesapeake in the early eighteenth century emphasized women's responsibilities. Eliza Smith highlighted recipes for cider, ale, and beer in *The Compleat Housewife* (1727), the first cookbook published in America. Recipe books that Chesapeake women created for themselves also included recipes for making alcoholic beverages. These "books" often were only twelve or so recipes, indicating that each was something that the woman felt she would need. For example, Martha Washington's cookbook included recipes "to make syder," mead, cherry wine, and elderberry wine.[12]

As greater numbers of Chesapeake men took control of alcohol production in the latter half of the eighteenth century, the cookbooks that women in the region purchased contained fewer recipes for alcoholic beverages. The most popular cookbook in the Chesapeake during the second half of the eighteenth century was Hannah Glasse's 1747 *The Art of Cookery, Made Plain and Easy*. Glasse shifted her recipes for brewing and winemaking to the back of the book in chapter 22. Distillations were pushed even further back, to chapter 25. William Ellis's popular 1750 *The Country Housewife's Family Companion* also repositioned alcoholic beverage recipes to the end of the book. Despite the fact that he marketed the book with the promise of "the several ways of making good malt; with directions for brewing good beer, ale, etc.," Ellis's book only included instructions for brewing on the last couple of pages.[13]

Chesapeake colonists read in Martha Bradley's 1770 *The British Housewife* the suggestion that brewing and cidering should be the work of men. "As we have given directions to the person who brews," she wrote, "to be careful in the choice of *his* malt and hops, we are here to give the same caution to the cider-maker, in the choice of *his* apples" [emphasis added]. Many Chesapeake colonists also owned Susannah Carter's 1772 *The Frugal Housewife*, a book that contained over five hundred recipes but none at all for cidering or brewing. And while Mary Cole told women at the end of *The Lady's Complete Guide* in 1791 that "malt liquors should not be passed over unnoticed, as the house-keeper cannot be said to be complete in her business, without a competent knowledge in the art of brewing," she meant only that women should be familiar with the process; she assumed that brewing would be performed by men.[14]

The first published cookbook written by an American, Amelia Simmons's 1796 *American Cookery*, also conspicuously shifted alcoholic beverage recipes out of the province of women. Simmons composed *American Cookery* in the spirit of post-revolutionary nationalism, making it the first cookbook to include recipes for cornmeal, pumpkin pudding, and another American novelty, spruce beer (molasses, yeast, and water with spruce boughs or needles for flavoring). The inclusion of spruce beer reflected a desire to showcase American foods more than any conviction regarding the propriety of women making alcohol, and the recipe appeared only at the final page of the book. Simmons's nationalism appears even more strongly in the 1800 edition of the book, when she added recipes for "Election Cake," "Independence Cake," and "Federal Pan Cake." *American Cookery* was widely printed as late as 1831 and was plagiarized repeatedly, appearing in 1805 as *New American Cookery*, in 1808 as *New England Cookery*, and in 1819 as *Domestic Cookery*. Simmons's influential cookbook, like other late eighteenth-century works popular in the Chesapeake, eliminated recipes for alcoholic drinks.[15]

At the same time that alcoholic beverage recipes disappeared from cookbooks popular in the Chesapeake in the latter half of the eighteenth century, they began to appear in husbandry books aimed at Chesapeake men with increasing frequency. Prior to this shift, seventeenth- and early-eighteenth-century husbandry books rarely included alcohol production recipes or instructions. For example, William Ellis's 1732 *The Practical Farmer* and George Cooke's 1741 *The Complete English Farmer*, both popular in the Chesapeake, never mentioned alcoholic bev-

erages. In the latter half of the eighteenth century, however, English works exhort-
ing men to take up distilling flooded the Chesapeake. George Smith's *A Compleat Body of Distilling* was a frontrunner. Originally published in 1725, the book
became extremely popular in the Chesapeake in the 1770s when the *Virginia
Gazette* bookstore in Williamsburg expanded and began stocking it. Smith taught
men to make elite women's alcoholic concoctions of aniseed water, angelica
water, and cinnamon water. In his widely read 1757 book *The Complete Distiller*,
Ambrose Cooper wrote that all forms of distilling were men's responsibility.
Cooper informed men that "distillation, tho' long practised, has not been carried
to the degree of perfection that might reasonably have been expected," and that
because female distillers had assumed that "the theory of distillation is very ab-
struse, and above the reach of common capacities," women had been "hardly sus-
pecting their art capable of improvements." Cooper urged men "to destroy this
idle opinion" and promised to teach "the distiller how he may proceed on rational
principles."[16]

Instructions for cidering and brewing aimed at men also began to appear in
other books in the Chesapeake. In *The Cyder-Maker's Instructor*, published in
Philadelphia in 1760 and well-read in the Chesapeake, Thomas Chapman told
men that cidering and brewing were their responsibility. He told men how to
make yeast, beer, raisin wine (raisins, yeast and water), and cider. His book, he
promised, "directs the grower to make his cider in the manner foreign wines are
made . . . [and] directs the brewer to fine his beer and ale in a short time." George
Watkins informed men in *The Compleat English Brewer* of 1768 that a man who
brewed carefully could expect to "equal the drink he meets with in the best
houses; probably to exceed it." Recipes and instructions even began appearing in
books intended for men that were not related to farm or household management
at all. For example, New Yorker Elijah Bemiss's 1806 *The Dyer's Companion*,
about the manufacture of dye and dyed cloth, included recipes for cider, apple
brandy, claret, gooseberry wine, raspberry wine, damson wine, grape wine, currant wine, strawberry wine, beer with and without malt, and molasses beer.[17]

By the end of the eighteenth century, books instructing men on cidering, distilling, and winemaking proliferated in the Chesapeake. John Richardson published *Theoretic Hints on Brewing Malt Liquors* and *Statistical Estimates of the
Materials of Brewing* in 1784. Alexander Morrice published *A Treatise on Brewing*
in 1802 "for the young brewers, and for the benefit of country gentlemen." Hundreds of books appeared in the latter half of the eighteenth and the early nineteenth centuries aimed at instructing men in brewing, distilling, winemaking,
and cidering. Many of these made an appearance in the Chesapeake, and soon

American men were writing and publishing their own alcohol instructional manuals. For example, Samuel M'Harry published *The Practical Distiller* in Harrisburg, Pennsylvania, in 1795; Samuel Child published *Every Man His Own Brewer* in Philadelphia in 1796; and Michael Kraffts composed *The American Distiller* in Philadelphia in 1804.[18]

More evidence of small planters understanding that distilling, even on a small scale, was now men's work appears in wills of Chesapeake men. While men in the seventeenth and early eighteenth centuries often bequeathed alcohol producing utensils to their daughters, they increasingly left such items to their sons in the mid-eighteenth century. Before the shift, for example, Bartholomew Andrews of Surry County, Virginia, left a still "to wife Elizabeth for life" in 1720. In the middle of the century, however, many men allowed wives and daughters to use stills during their lifetimes, then required them to be passed to a son once the women had died. In 1746 Thomas Haynes in Prince George County left his son "one hot still for brandy" once his wife's use ended with her death. In 1750, William Walker similarly left his wife Jane "the still" with the provision that their sons would inherit it after she passed away. Later in the century, men skipped over daughters and wives entirely and bequeathed stills directly to sons. In 1750, Joseph Carter of Spotsylvania County, Virginia, was survived by his wife, four sons, and three daughters. His will left his brandy still to his sons and left no equipment for making alcohol to his daughters and wife. Examples of fathers leaving stills directly to sons in the Chesapeake explode after 1760. For example, Tobias Purcell left his still to his step-son in 1761, and Matthew Harrison left a still to his son Benjamin in 1764.[19]

New inventions made alcohol production easier and cheaper, and permitted men of all means to take charge of making alcohol. The new scientific instruments of the alembic still, the thermometer, saccharometer, and hydrometer, as well as the new instruction manuals, meant one did not need to be an intuitive master of alcohol production. With these tools, unskilled laborers could now work in household and commercial breweries and distilleries. Writers were quick to point out the advantage of cheap labor that the new tools offered: "If you want . . . to employ yourself on some other business, having one of these thermometers, you need not stand at the side of the copper to watch it, but . . . leave anyone of your ordinary workmen to take your liquor, and turn over when the quicksilver rises to the index: this will save the brewer a great deal of trouble," one author advised. In fact, the thermometer and saccharometer made it so easy to make alcoholic beverages that another author recommended that brewers purchase "blind thermometers" in which the scale could be hidden in the brewer's or distiller's pocket so that his workers would not learn his methods and be able to found businesses of their own.[20]

The invention of the alembic still, or side distilling, in particular, made the process easier. Side distilling became known in England around 1720, but it was not practiced in the Chesapeake until the 1760s. Before the invention of side distilling, stills were very large and expensive pieces of equipment, and distilling was a complex process. In the side distillation apparatus a coil connecting two containers was immersed in a basin of cold water so that the alcoholic vapors condensed more rapidly. Since the steam did not need to travel up a lengthy rise or drop, the parts of the still were smaller, easier to transport, and less prone to breakage. The process and construction was simple and effective enough that home distillers continue to use side distillation today. The apparatus spread throughout Europe; the French, in particular, became enraptured with distilling. There, Antoine Parmentier constructed the first modern still, where the raw materials to be distilled were heated in a boiler over an oven fire, producing a cheaper liquor that was less tainted by impurities. The subsequent popularity of French distilled liquors in England led English leaders to encourage the local production of distilled liquors — an encouragement they would later regret, as thousands of people became addicted to gin in the early nineteenth century.[21]

Popular books taught Englishmen how to build the new stills starting in the 1720s. One was George Smith's *The Complete Body of Distilling* (1725), which demonstrated how to build a simple three-gallon alembic still. Smith's extremely popular work went through at least eleven editions by 1813. Smith explained that in the seventeenth and early eighteenth centuries, distilling had remained the province of the very wealthy. Distilling had required a space large enough to contain the still, worm-tub, and pump in a row, with an inclined and paved floor and a chimney. Smith's three-gallon still, in contrast, was small enough to fit almost anywhere and required little infrastructure. Smith's still was further improved upon in Ambrose Cooper's *The Complete Distiller* in 1757, which not only detailed how to build and use an alembic but was also the first English work to give explicit instructions on distilling rum. Cooper's work was popular, with at least five editions by 1810. The alembic still, as Cooper explained, required two containers and a worm and, according to Cooper, was "one of the most speedy and profitable [stills], as it required fewer preparative[s], and less time." Smith's, Cooper's, and similar publications taught households in England how to build their own rudimentary stills and would spread to the Chesapeake in the 1760s.[22]

Improvements in distilling technology continued to make distilling easier for Chesapeake residents throughout the nineteenth century. In 1801, Alexander

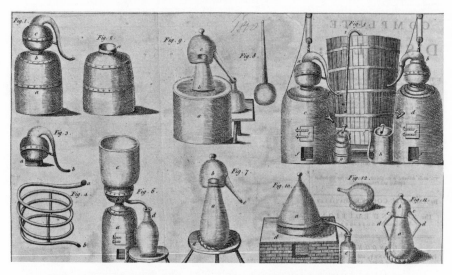

Men in the Chesapeake began reading about, building, and purchasing
the new alembic stills during the latter half of the eighteenth century.
Ambrose Cooper, *The Complete Distiller,* 2nd ed. (London: 1760). Image
used with permission of Special Collections, John D. Rockefeller, Jr.
Library, the Colonial Williamsburg Foundation.

Anderson patented a condenser that heated wash more quickly. Henry Witmer
soon patented an improvement on Anderson's condenser. Anderson's condenser
with Witmer's improvement commonly allowed a 110-gallon still to be run off
eight or nine times in twenty-four hours, a significant improvement over the three
runs that had been typical before. Not only could the new stills produce larger
amounts of alcohol more quickly, they also produced more alcohol from the same
amount of wash. Authors raved about the new technology, assuring the distilling
public that with the new stills, one bushel of grain could produce three gallons of
spirits. "Accordingly," noted authors, "we find men of science, men of capital,
lawyers, doctors, and merchants abandoning other pursuits to learn the art of ex-
tracting spirit from grain."[23]

Other inventions soon followed, like "Montgolfier's water raiser" for hydraulic-
powered stills, which allowed a distiller to eliminate the cost of hiring a person or
a horse to pump the water, and mechanical stirrers, to reduce the human labor
necessary to stir the materials. By 1804, these inventions led Michael Krafft to
write in *The American Distiller* that stills had been improved so much that "no fu-
ture period can boast that we have left them the smallest shadow of improvement"
in distilling technology. From January 29, 1791, to April 25, 1812, the U.S. Patent

Office registered sixty-eight new distilling patents. These patents were only part of the tremendous interest in distilling. Tools like the new small stills meant that the level of skill formerly necessary was no longer required for distilling.[24]

Small-planter households not only read about the new distilling techniques, but they also created or purchased new-style stills and used them to distill cheap molasses into rum. During the latter half of the eighteenth century, stills became simpler, easier to transport, less expensive, and more common. When Benjamin Bramham wanted to sell a still in 1769, he noted that it weighed only 37 lbs. By the 1760s, colonists could buy ready-made stills or the components for stills in Virginia. Advertisements in the *Virginia Gazette* indicate that by the 1770s, still capacities in the Chesapeake ranged from the small 30-gallon still to the much larger 400-gallon still. Robert Lyon sold "a large assortment of copper, pewter and tin ware" at his store in Williamsburg, while John Greenhow, another Williamsburg merchant, sold "most sorts of pewter, tin, copper," as well as wire and all sorts of cast iron. Kidd & Kendale advertised "still-worms made and mended" in Williamsburg beginning in 1769. In 1772, James Haldane of Norfolk advertised that he made all sorts of copper and brass work, including stills and brewing coppers "at the most reasonable rates" "for cash or country produce." Increases in coastwise shipping allowed Chesapeake colonists to get stills or their components from Philadelphia and New York, where merchants competed for this business. Stills, worms, and still heads ranging in value from £1 to as high as £20 became common in estate inventories in the last third of the eighteenth century. One study found that between 1780 and 1800 25 percent of households in Augusta County, Virginia, owned stills.[25]

Colonists used their stills to make rum from molasses, which became cheaper over the course of the 1760s. In 1764, the British government lowered the duty on molasses imported from the French and Spanish Caribbean islands from a prohibitive six pence per gallon to three pence per gallon. In 1766 the molasses duty dropped again, this time to only one penny per gallon on all molasses entering the colonies. In addition, during the second half of the eighteenth century, West Indies planters expanded their sugar production. The resulting oversupply further decreased the price of molasses and rum. From 1737 to 1742, Virginians imported an average of 16,659 gallons of molasses annually; from 1742 to 1769, they increased their average molasses imports to 40,054 gallons of molasses annually. Only small amounts of the thousands of gallons of molasses imported went into household cooking, while the vast majority went to making rum. Colonists also used the cheap molasses to make their own unhopped molasses "beer," which required fermented molasses and water.[26]

This picture of a late-eighteenth-century still indicates that stills became even smaller, simpler, and less expensive at the end of the century. Image reproduced by permission of the Colonial Williamsburg Foundation.

Books on the new distilling methods sold briskly in the Chesapeake after 1760, when the region saw its first bookstore. The *Virginia Gazette* bookstore and printing office advertised Smith's *Complete Body of Distilling* at least twenty-one times between 1770 and 1776. In total, the *Virginia Gazette* bookstore offered 273 works on science and 26 titles related to agriculture, many with distilling instructions, including Robert Maxwell's *Practical Husbandman*, Arthur Young's *Farmer's Guide*, and Duhamel du Monceau's *Elements of Agriculture*. The store also sold at least eleven encyclopedias with instructions on side distilling and a variety of periodicals containing alcohol production advice, including *Gentleman's Magazine, London Magazine, Monthly Review, The Guardian, The Connoisseur,* and *American Magazine.* The instructions in these works offered to democratize distilling, just as the title of one proclaimed: *Every Man His Own Distiller.*[27]

Although many small planters could not afford books about the new distilling techniques, instructions on side distilling began to fill the pages of the cheaper almanacs and newspapers. The *Virginia Almanac*, which began in the 1750s, expanded in the 1760s to include articles on distilling. For instance, William Rind included instructions for distilling in his *Virginia Almanac* for 1761. Benjamin Franklin published an article in his 1765 Philadelphia *Poor Richard's Almanac,* popular in the Chesapeake, on "how to manage the distilling a spirit from rye." Franklin also included instructions on brewing from corn. The *Virginia Gazette* advertised almanacs that included alcohol production techniques, like one with instructions for "an wholesome liquor made from Indian corn" in 1761. A *Virginia Almanac* of 1770 included a lengthy article "upon the distillation of persimmons, communicated to the American Philosophical Society," while another 1770 *Virginia Almanac* included "directions for making cider in the manner foreign wines are made." David Rittenhouse's *The Virginia Almanack for the Year of our Lord God 1776* reprinted alcohol instructions from Malachy Postlethwayt's *Dictionary of Trade and Commerce* and also discussed alembic stills. John Skinner's *The American Farmer,* a weekly newspaper begun in 1819, gave advice on making persimmon beer, ginger beer, cider, brown spruce beer, white spruce beer, and managing fruit trees. John Taylor of Caroline published a series of agricultural essays in a Georgetown newspaper in 1803 that included directions for making cider and distilling beverages, a series that he expanded into the book *Arator* in 1813. Most colonists who were literate read almanacs and newspapers, and both were read aloud in taverns for those who could not read, so numerous colonists had the chance to learn about the new stills and distilling techniques beginning in the 1760s.[28]

For men who remained hesitant to assume women's work, scientific and agricultural societies also taught the new alcoholic beverage production techniques to men. In 1769 the American Philosophical Society began a series of weekly meetings for the "mutual improvement in useful knowledge" aimed at elite men in Philadelphia, and it published many of its members' papers in the *American Magazine.* By July of 1773, a group of one hundred Virginians had established a similar society called the Virginia Society for Advancing Useful Knowledge. The meetings and publications of these societies helped to spread alcoholic beverage experimentation among men. For example, Isaac Bartram, a fellow of the American Philosophical Society, presented a paper on the distillation of persimmons and urged farmers to cultivate persimmons for alcoholic beverages. Virginia planter John Mercer's papers include a circular from the Maryland Farmers' Club stating that the club had organized in order to "supply the means and the opportunity for the diffusion of useful knowledge and discovery." Agricultural societies traveled to give demonstrations to men who could not attend meetings, as, for example, John Skinner announced in *The American Farmer* that the Maryland Agricultural Society at Easton was giving traveling demonstrations. In the latter half of the eighteenth century, men in the Chesapeake increasingly taught their fellows how to make alcoholic beverages, both in print and in person.[29]

Why did men in the Chesapeake need publications and agricultural societies to learn how to make alcohol? Why did they not just ask their wives? The products they were making, especially brandies and rum, were unknown to Chesapeake women, and they also used technology, including alembic stills, that were also unknown to women. The inspiration and source for Chesapeake men was not their wives but the gentlemen scientists of England, who were themselves imitating men making distilled spirits in Italy and France.

Cidering was becoming much easier as well, allowing even families of small means to make much more cider. Seventeenth- and early-eighteenth-century cider presses like John Worlidge's "ingenio for the grinding of apples" had been expensive and hard to obtain. Worlidge's press, which squashed the apples between rollers and then strained the pulp through a hair bag sieve or flannel cloth, cost £10, a prohibitive price in the seventeenth century.[30]

In the latter half of the eighteenth century, small planters began making cider more efficiently. Soon almanacs began including instructions from books like Worlidge's (written to advertise his press), and homemade cider troughs and presses became common in estate inventories. The drop in the cost of cider presses is particularly evident in probate assessments. In 1744, Capt. Mathew Kenner's "2 apple mills cribs and platforms" was assessed at £3. By 1777, William Baily's "1 apple mill and press, tubs, flat forms and troughs," was valued at only £1 5s. In the latter eighteenth century, homemade cider troughs and presses were usually assessed at only £1.[31]

Experimentation with grafting introduced a new kind of apple to the region, which also increased small planters' production of cider and brandy. Until the mid-eighteenth century, Chesapeake colonists made their cider mostly from the Cattaline apple, which made a sour and ephemeral cider; or they used the small, bitter, and wormy local crab apples. The introduction of the Hewes (sometimes spelled Hughes) crab apple to the region in the mid-eighteenth century allowed planters to produce a sweeter, slightly cinnamon-tasting cider that lasted longer. In 1774 one northern Revolutionary War soldier sent to Virginia noted in his diary that "Hewe's crab-apple is much cultivated in Virginia. I have tasted better cider made of it than any I ever drank made from northern fruit. The cider is quite pale and clear, but of most exquisite flavor. 'Tis certainly worthy taking much pains to propagate these trees with us." The higher sugar content of the Hewes crab apple hastened and increased fermentation. Moreover, the Hewes crab apple was a sturdier apple, less susceptible to worm infestations and disease. It was a vigorous grower, which helped its trees spread. It was simpler and faster to press the Hewes crab apple than previous apples because, as the author of an early-nineteenth-century manual about cidering pointed out, it did not need to be pressed through a sieve. The apple itself was "sufficiently fibrous and tough" to provide its own filter. The resulting juice was "white, and clear as spirit from a still, without any mixture of pulp," "making a high flavored sprightly liquor, requiring but little fermentation, and easily fined." Some planters selling orchard land stressed that the property contained Hewes crab apples. "To be sold," John Fox advertised in 1772, "a valuable tract of land in Gloucester," with "an orchard of Hughe's apples, and several other choice fruits." The other fruits Fox mentioned probably included the Taliafero, Roan, and Gloucester White apple varieties, all introduced in the mid-eighteenth century and all making a cider that was better than the original Cattaline apple cider. With the new alembic stills, colonists could distill this improved cider into non-spoiling apple brandy.[32]

One way the new apple varieties spread was through the new commercial botanical trade, which let small-planter households obtain the same types and diversity of fruit as their wealthier neighbors. In the seventeenth and early eighteenth centuries, planters who wanted orchards had to hire grafters. In the second half of the eighteenth century, colonists could increasingly order trees from new commercial nurseries. William Smith advertised in the *Virginia Gazette* in 1755 that he had forty-six types of fruit trees for sale; Christian Lenman sold "a quantity of very fine young apple-trees, both grafted and ungrafted . . . all in an excellent thriving condition and fit to plant out this ensuing spring" as well. "Any gentlemen that send their orders," advertised nurseryman Thomas Sorsby, "may depend on being as punctually served as if they were present." William Prince's nursery in Long Island had a thriving catalog business by 1771. Prince sold 120 trees to George St. Tucker of Petersburg, Virginia, who then distributed scions and fruit pits from the new varieties to his friends and customers. Small planters could also learn how to graft their own fruit as articles on the topic appeared in newspapers and almanacs in the second half of the century. For instance, the 1769 *Virginia Almanac* included instructions on "Dr. Dimsdale's directions for inoculation" of fruit trees.[33]

As a result of the diffusion of the new apples, presses, and stills, travelers reported that even poor planters could make fruit ciders and brandies by the latter decades of the eighteenth century. Middling planter William McClemmey of Manakin, Virginia, left a still and worm to his two sons in 1750, for example; and Isaac Handy, a middling merchant, owned a still and one hundred gallons of cider to distill in it when he died in 1762. In 1791 a Frenchman traveling in Winchester,

Virginia, noted a poor German blacksmith who distilled and sold "whiskey," meaning any distilled drink. (One traveler explained that "rum is distilled, which is a kind of whiskey.") At the home of another poorer-sort family, the Frenchman noted, "I drank some old whiskey, distilled on his place." Distilled drinks had become so popular that J. P. Brissot de Warville noted, while traveling through Blandensburg, Virginia, that "we found nothing to drink except eau de vie [brandy or spirits] or rum and water." Even "the very meanest and hilly land are proper for peach trees," commented one visitor, "every planter, almost having an orchard of these trees. The brandy made from that fruit I think is excellent and they [make] it in general in sufficient quantities." By the mid-eighteenth century, all moderate farms had at least a small apple orchard, and many also had a peach orchard. In 1785 traveler John Joyce wrote to his uncle that in Virginia "the drink chiefly used in this colony it is generally cider, every planter having an orchard and they make from 1000 to 5 or 6000 [gallons] according to their rank and fortune."[34]

The American Revolution brings into sharpest relief the transfer of alcohol production from women's hands to men's. Halfway through the war, George Washington and the Quartermaster's Department adopted rum and whiskey as part of the official ration. At the same time, they barred women from selling alcoholic beverages to the army as a way of reducing the number of people traveling with the army as camp followers. The result was that men had to make the alcohol. Thus the formal transition of labor that had begun in the scientific societies continued informally in the camps.

Alcohol was essential for soldiers. Alcoholic beverages provided much-needed calories for soldiers, made the spoiled food somewhat more tolerable, offered an alternative to disease-ridden water, and supplied a sense of warmth during the numerous clothing shortages. Constant shortages of salt meant that much of the food that soldiers received was rotting or spoiled — when they received food at all. At Morristown, men often went without meat and bread for a week at a time. At Valley Forge, one officer recorded a recipe for cooking spoiled pork and hog fodder. The only utensil issued to troops was a camp kettle: one kettle for every six men. Private Elijah Fisher recorded during a campaign near White Marsh, Pennsylvania, in December of 1777, that the lack of utensils required anyone cooking meat "to throw it on the coals and broil it." Wood shortages also made cooking difficult and sometimes required soldiers to eat their meat or fish raw. Soldiers on active operations were supposed to receive hard bread, but it was frequently unavailable, and soldiers learned how to mix flour with water and "cook" it on hot stones into dirty sodden cakes. As Fisher explained, "The water we had to drink and to mix our flour with was out of a brook that run along the camp, and so many a dipping and washing it which made it very dirty and muddy."[35]

Alcoholic beverages not only improved the food but were also thought to promote health and reduce fatigue. Army doctors frequently prescribed two or three bottles of wine per sufferer per day to cure "putrid fevers." Henry Knox urged the commissary to give the soldiers rum because "we have found by experience that this would support the men through every difficulty." "The lives of our men," George Washington reminded Quartermaster Robert Morris, "depends upon a liberal use of spirits in the judgment of the most skilful physicians." Alcohol consumption rose further because of fears of water drinking. For instance, in 1780 Captain George Fleming wrote to his superior that "I have been unfortunate in losing Peter Young, by his taking a hearty draught of cold water."[36]

Granting extra alcohol rations was one of the few ways that army authorities could urge men to fight and reward them for their work. Potential soldiers were recruited with alcoholic beverages. Colonial militia leaders provided alcohol after drills to persuade men to join, and the resulting soldiers expected the additional drinks to continue after drills, battles, and on special occasions. On St. Patrick's Day, for instance, officers in the Pennsylvania Line drew a quart of rum each. On July 3, 1777, General Lachlan McIntosh gave his Savannah troops an extra quarter cask of rum "to celebrate the anniversary of the most extraordinary and glorious revolution in the history of mankind." "I felt very unwell, this whole day," soldiers frequently noted in their journals, "from last night's carouse."[37]

Soldiers were given extra alcoholic beverage rations before battle. They typically also received a gill (about four ounces) of rum before marching into a fight. Soldiers received double rations of rum during sieges or in cold or wet weather and for any task, such as digging trenches, that was deemed extraordinary service. For example, on March 15, 1781, Nathanael Greene ordered that the soldiers each be given an extra gill of rum at breakfast to prepare for what would come to be known as the Battle of Guilford Courthouse. And when the Southern Army marched to the Battle at Eutaw Springs, Colonel Otho Holland Williams recorded that "we halted, and took a little of that liquid which is not unnecessary to exhilarate the animal spirits upon such occasions."[38]

Until 1781 the official ration during the war included "1 quart of spruce beer or cider per man per day, or nine gallons of molasses per company of 100 men per week," with the molasses to be brewed into molasses beer. The Continental Congress encouraged the army's men to obtain their food the way English armies had traditionally done: by living off the land or by purchasing from licensed sutlers or local residents. Congress assumed that the traditional methods would continue to suffice, and it published lists of the supplies the army needed, including beer and cider, as a way of encouraging sutlers to attend to the army. An example of local procurement occurred in the summer of 1775 when Richard Backhouse supplied Thompson's Pennsylvania Rifle Regiment with small purchases of beef from local farmer John Hendershot, mutton from Ann Snook, and bread from Jane Allen. Elizabeth Beard of Campbell County, Virginia, submitted a petition to be repaid for the nine diets (complete rations) that she had provided for the North Carolina Light Dragoons in February of 1779. Widow Agnes Jones submitted a petition for providing 450 pounds of beef to the army, while William Arthur asked to be repaid for thirty-eight gallons of brandy. Captain Rogers recorded in March of 1777 that he had purchased two bottles of wine, supper for six men, and one sheet from Elizabeth Wilson in Maryland.[39]

The army ration from 1775 to 1778 called for simple brews — spruce beer, molasses beer, and cider — that had long been women's province. The sources of this alcohol were camp followers and sutlers. Camp followers were the wives, children, and prostitutes who followed and supplied the army to make money, assist their husbands, and support the revolution. These women washed, sewed, cooked, and brewed for the troops and nursed them when they were sick and injured. Women had long played a valuable role in provisioning the English and colonial armies and were proud of their work. For example, Martha May stressed her commitment to the army when she wrote to Henry Bouquet in 1758, "I have been a wife 22 years to have traveled with my husband every place or country the company marched to and have worked very hard ever since I was in the army." When Mary Cockron applied for a pension in 1837 for her own and her husband's service to the Continental Army, she stated that she "drew her rations as other soldiers did."[40]

At the beginning of the Revolution, Congress and army officials tried to ban soldiers from drinking rum, arguing that cider, ale, and beer were healthier and caused less drunkenness. For example, George Washington twice attempted to bar purveyors of distilled liquor. In reaction, soldiers stole liquor, sold their clothes and other items to purchase liquor, and rebelled when they did not receive liquor rations. Washington then tried instituting price controls and punishing sutlers for selling liquor to soldiers. Typical punishment for a sutler found selling rum to a soldier was two hundred lashes and forfeiture of the liquor. Soldiers who were discovered drinking rum or were found drunk had their alcohol rations withheld and were frequently court-martialed and whipped for their offenses.[41]

Congress and army officials found it difficult to keep the army supplied, to provide ale or beer for the solders, or give them the necessary ingredients to make the drinks. Molasses, yeast, malt, and hops were continually in short supply. "I find no malt yet," supply agents frequently informed the Commissary General of Provisions, Joseph Trumbull. When Trumbull asked agents to buy hops, they reported that they, too, were scarce. Congress tried to provide beer and cider by decree, such as when it told the Board of War on July 25, 1777, to contract for a supply of beer and cider. As usual, George Washington noted ten days later that no such supply had been procured. Washington was forced to conclude that "no army was ever worse supplied than ours with many essential articles of it. . . . Neither have they been provided with proper drink. Beer or cider seldom comes within the verge of the camp." Consequently, alcoholic beverage rations became essentially anything the soldiers or officers could locate, be it cider, ale, beer, wine, rum, or whiskey, which they usually obtained from local men and women.[42]

George Washington, Robert Morris, and Congress were working to reduce the role of women in the army, particularly their role as sutlers. In April of 1778, Ephraim Blaine, the Deputy Commissary General of Purchase of the Middle Department, approved George Washington's request that the soldiers be issued "1/2 gill of rum or whisky per day in lieu of beer." Not only did the army reverse its disapproval of the use of rum and whiskey, it then turned to men, and men alone, to provide the new rations. In 1781, Congress instituted a system in which Congress would offer contracts to individual men to provide a complete set of rations to a particular section of the army for one year. The specific daily ration to be supplied included one pound of bread, one pound of beef or three-quarters of a pound of pork, and one gill of rum per man per day, as well as one quart of salt, two quarts of vinegar, eight pounds of soap, and three pounds of candles for every hundred rations.

Congress and the state assemblies also took the opportunity to pass restrictions against women accompanying the troops. General Braddock permitted Virginia and Maryland troops only six women for the Regiments and Independent companies; five women to the Light Horse, seamen, and artillery; and four women to the carpenters of the Rangers. In contrast, a British account of March 1779 shows more than 1,550 women and 968 children traveling with 4,000 British soldiers. Congress also barred sutlers and local men and women from selling alcoholic drinks to soldiers. These actions combined to give the business of supplying the Continental Army's alcohol to men alone.[43]

Other factors in the army's new system favored men. The calls for proposals for contractors were advertised in newspapers, at a time when the majority of the lower sort, minorities, and women were illiterate. Moreover, it was impossible for a married woman to sign a contract, since all *femes coverts*, or married women, were considered legally dead and thus were ineligible to engage in contracts. Merchants who could provide the entire ration were deemed preferable to individuals who could only provide partial rations. It became increasingly important to Morris and others that the contractors have excellent reputations and letters of credit built from years of business. When Morris offered to write a letter on behalf of Baltimore merchants Matthew Ridley and Mark Pringle for obtaining European goods for the army, he wrote, "I am most perfectly satisfied of your honour, integrity and solidity [and] I very readily agree to guarantee the payment of any bills which your Matthew Ridley esq. shall give." [44]

Finally, the new contracts called for rum, not the traditionally female-produced cider, spruce beer, or molasses beer. There were good reasons for the army to switch to rum. Distilled liquors were less likely to spoil, required less space since they contained greater alcohol by volume, and saved the grains used in brewing beer for flour and bread. However, the decision to use rum as staple issue clearly favored men.

The effective favoritism shown to men by Congress and the Quartermaster Department was not unintentional. It was the culmination of a century of concern about the role of women and armies. In the mid-seventeenth century, some Englishmen attempted to prevent female brewers from joining the army because of fears of women distracting men. For example, Lord Bridgewater requested in 1641 that the constable replace his troops' female brewer by "find[ing] out a man to do it." During the colonial wars, Americans noted British camp-follower practices and their ambivalence toward women. Women were needed and tolerated by the British army to wash clothes, brew beer, and make soap, but they were also inspected for venereal disease and often were drummed out of camp. During the Seven Years War, the concern about women traveling with the army expressed by Lord Bridgewater in 1641 grew among the leaders of the American provincial troops. In 1780 Washington ordered women who were not wives to leave the camps, and in 1781 he purposely adopted the contract system that favored men and a rum ration made by men to reduce women's roles further. Thus war itself cemented the transition of alcohol production and provision from women to men.

George Washington would have liked to remove women from the army entirely because he felt that women slowed the army's movements and made it look less professional. But he knew that he would lose husbands and fathers if he did not allow women to follow and supply the army with rations. However, he made certain that women did not feel welcome in the army. When the army marched on Yorktown, Washington ordered the troops to deposit both their baggage and their women at West Point so as not to slow the army's progress. He issued similar orders lumping women with baggage on other marches. Army officials also required women who washed and brewed for the troops to undergo regular examinations for venereal disease. For example, on July 1, 1777, the commander of a Delaware regiment ordered "that the women belonging to the regiment be paraded tomorrow morning and to undergo an examination [for venereal disease]. . . . All those that do not attend to be immediately drummed out of the regiment."[45]

The army's transition to alcohol made by men cemented the idea that in the new American republic making alcohol was men's work. And by the late eighteenth century, the diffusion of the new apples, presses, and stills meant that even men of small-planter status could make most of their alcohol their households required at home. Individual households still continued to run out of what they wanted, and not every household made each kind of liquor. Small-planter households still needed to shop or trade for some alcoholic beverages. However, it is interesting that in the late eighteenth century, when small planter households needed alcoholic beverages, they eschewed trading with the large planters, instead developing a trade with each other and occasionally purchasing rum at the increasing array of stores and distilleries.

Francis Taylor and his father, small to middling planters in Virginia, provide an example of the neighborhood trade among men (only) of like status. Francis Taylor (1747–1799) was a Revolutionary War officer. His diaries (a generous term, since the entries are terse and sporadic) commence after the war ends and record Taylor's day-to-day-life in Orange County, Virginia, from 1786 to 1799. Although his ancestors were wealthy, property divisions over generations meant that his inheritance was modest. "Midland" comprised four hundred acres of land and a two-story frame house of twenty-five by twenty-three feet with one room on each level. Like a typical small to middling farmer, Francis Taylor owned one slave and four horses, and had no cattle or servants.

What is particularly striking in the Taylor diary is the impression that men like Francis Taylor no longer depended on large planters for their alcoholic beverages or supplies. By the latter half of the eighteenth century, small planters were sharing the tasks of the alcoholic beverage trade with each other: exchanging advice, ingredients, tools, and drinks.[46]

The shift began with advice. "Saw J[oseph] Clark," Taylor recorded in August of 1792, "who says he thinks it will be worth getting peaches to make brandy." Taylor checked this suggestion with his father, who agreed that additional peaches would be necessary. "He says he does not think there will be enough to make brandy," noted Taylor. In April of 1788, Taylor took Major Moore's advice and grafted some of his pear trees with Moore's pear stocks. Exchanging graftings to encourage the variety and hardiness of fruit was another way that small planters expanded alcohol production through small planter and kin exchange. In March of 1790, Taylor "sent to Mr. Ingram's and got some young trees out of his apple orchard." In October of 1795, Taylor gave extended relative and neighbor Joseph Taylor some peach tree graftings. Likewise, when Joseph Ball wanted to move his peach trees to fresher soil, he wrote to his friend for "some help from the other plantations with carts and men."[47]

Small planters stayed aware of what alcohol their neighbors were producing, which was important if they wanted to observe the process, ask for advice, offer suggestions, or purchase the product. For example, Francis Taylor knew that Hubbard Taylor was cidering and that Joseph Taylor had "sent his cart to H[ubbard] Taylor for [a] cask of cider." This same cart in turn brought thirty-five gallons of cider for George Taylor. Francis Taylor's father then bottled the cider that Hubbard Taylor had sent. In April of 1789, Francis Taylor went to Thomas Jones's store and found that Jones had obtained a quart of brandy from George Taylor. In September of 1789, Joseph Taylor sent for and received a jug of Francis Taylor's brandy.[48]

Neighbors sometimes shared the labor of producing alcohol. "Mr. Shepherd came in the evening," wrote Francis Taylor, and "brewed persimmon beers." In October of 1786, Francis Taylor beat cider at Hubbard Taylor's house. On another occasion, Francis Taylor remarked that George Taylor had gone to Joseph Clark's farm and that the two men had made brandy together, each taking a share home. Joseph Taylor sent apples to Francis Taylor's plantation several times to be beaten into cider. Neighbors also assisted each other with obtaining ingredients. For example, in September of 1796, Joseph Taylor borrowed a cask from Francis Taylor. When he returned it, he brought Francis seventeen gallons of brandy from Major Daniel. In December of 1786, Francis Taylor went to town and paid shopkeeper May Lee for the two gallons of molasses that Joseph Taylor had brought from the store to make Francis Taylor's molasses beer a week earlier.[49]

Small planters and kin also loaned and borrowed the materials necessary for alcoholic beverage production. Francis Taylor "boiled persimmon beer and put [it] up with hops and yeast in a cask borrowed of Capt. Burley" in February of 1790. He borrowed apples from Charles Taylor in October of 1786 and again in September of 1790. Robert Taylor borrowed Francis's father's still to make apple brandy in September of 1792; while in May of 1792, Francis Taylor returned a gallon pot to Mr. Howard that Taylor's father had borrowed to stew apples.[50]

Finally, small planters and their kin exchanged the products of their labor. For example, Francis sent a slave with an empty keg to Mr. Graves to obtain eight gal-

lons of brandy in April of 1789. On the same day Thomas Barbour stopped by
Francis Taylor's house and asked if he wanted a cask of cider; and in July of 1789,
Francis Taylor gave some of his brandy to Charles Taylor in exchange for whiskey.
In November of 1789, Francis Taylor exchanged brandy for rum with Major Lees,
mirroring a similar exchange made by Taylor's father four months later.[51]

While all this neighborly sharing of the labor and products of cidering and
distilling helped to build good feeling and small planter and kin relationships, at
its foundation it was economic. Francis Taylor recorded these transactions because he expected to be repaid. "B[enjamin] Taylors letter mentioned [sending]
three gallons," of alcohol, noted Francis Taylor, "but measured only two [and] 5/8
gallons." In another example, Francis Taylor recorded that Joseph Taylor borrowed "about two gallons" of brandy from him, while in May of 1788, Francis recorded that he had "paid Hansford for the cider I had last week." When Reuben
Taylor, Richard Cowthorn, and Joseph Langham called on a seemingly neighborly visit, their central purpose was to obtain the "two quarts brandy my father
owed Langham."[52]

Sarah Fouace Nourse's brief diary, kept sporadically from 1781 to 1783, offers
similar evidence of small-planter alcohol exchanges. Sarah and her husband,
James, lived in Berkeley County, Virginia. At the time that Sarah Nourse's diary
begins, James Nourse, who had spent fifteen years as a woolen draper in London
before immigrating to Virginia in 1754, was at least in his sixties. Sarah Nourse
opened her diary by noting "Mr. Nourse brewing" in April of 1781. James Nourse
later bought a still from his neighbor, Mr. Briscow, in May of 1781. Before he could
buy the still, according to the diary, James Nourse visited his neighbors to "raise
the money for the purchase" and collected what they owed him from previous exchanges. James Nourse then engaged in distilling with a person that Sarah called
"the stiller." Distilling, or "the stiller," evidently did not suit Nourse, because by
October Sarah noted that her husband and "the stiller" had "agreed to part."

The Nourses relied on small planters and kin to sell their alcohol for them,
perhaps because their age made travel difficult. Sarah noted on one such occasion that "Bob returned — made but a middling sale of the beer." "Bob and Jack,"
she wrote on another occasion, "gone early with beer for sale to prisoners at Winchester," but the Nourses were again disappointed with their sales. Another time
when a neighbor from Winchester came with a wagon to pick up liquor, he
"brought no money," and Sarah declined to give him anything. Neighbors they
might be, but Sarah Nourse expected to be paid.[53]

The journal that Colonel James Gordon of Lancaster County, Virginia, kept
from 1758 to 1768 provides further evidence of men assuming the alcohol trade.

This drawing, from a manual advising women on how to stock a pantry, indicates some of the tasks women had to complete for the household and suggests why they did not resist men's claims to alcoholic beverage production. Hannah Woolley, *The Queene-Like Closet* (London, 1681), reproduced by permission of the Huntington Library, San Marino, California.

Like Francis Taylor, Gordon knew what his neighbors were concocting. "Robert Hening began to still whiskey," Gordon recorded in April of 1758, "which I believe will answer very well." In July, Gordon knew that Colonel Conway was selling brandy, and he traveled to Conway's residence to purchase three gallons of it. Gordon himself focused on cider, reporting regularly that he was "very busy with our cider" and recording his desire to make five hundred gallons of cider in the fall of 1758. Like Taylor and Nourse, Gordon exchanged farming and brewing methods with his family and neighbors, for example, when he "went out with my brother to see his farm, which is very well managed." Gordon also exchanged recipes and advice with neighbors: "Sent to [the] mill for meal to make brandy," Gordon noted in his journal, "according to Mr. Criswell's directions."[54]

Despite the shifts, even following the Revolution, cidering, distilling, and brewing in the Chesapeake was less advanced than in England. By the late 1780s, almost all London breweries employed steam engines, powerful pumping systems, and mechanical mashing rakes. By 1800, commercial alcoholic beverage producers in urban England no longer needed to hire men or horses to carry, rake, or grind grain because they had machines to do this for them. In contrast, most Chesapeake men either made alcoholic beverages by hand or oversaw their slaves making alcohol by hand. In the Chesapeake, as in the rest of the West, alcoholic beverage production was no longer women's province.

Chesapeake women rarely resisted this change. In fact, women supported the transition to an exclusively male concept of alcohol production when they purchased the alcohol that men had produced. It is possible that early American women wanted to masculinize alcoholic beverage production as much as did anyone. Since women gained little by making alcohol in the early Chesapeake, they might have been just as happy to forget that making alcohol had once been women's work.

CHAPTER SEVEN

"He is Much Addicted to Strong Drinke"

The Problem of Alcohol

In 1660, Robert Warren sat drinking in a Northampton County, Virginia, courtroom, his voice growing louder and louder. He began "rudely intruding" on the court proceedings, "interrupting" the judges and "upbraiding" them for their judgments. The judges' response was to ask the sheriff to remove Warren from the courtroom, and he continued drinking outside. No one appears to have chided Warren for drinking to drunkenness, and no one threatened any punishment for his public inebriation. In the seventeenth and early eighteenth centuries, Warren could have continued his drinking outside with anyone he wanted, black or white, and no one would have been likely to bother him. After the mid-eighteenth century, however, his drinking escapades would have earned him censure. This chapter focuses on how drinking came to be seen as a problem in the Chesapeake during the latter half of the eighteenth century.[1]

Concerns in the Chesapeake about alcohol, and particularly about drunkenness, began to arise in the mid-eighteenth century and focused on drinking by slaves and servants. It is impossible to know what anyone thought of elite women's alcohol consumption, because no one left any comments on the topic. The lack of commentary suggests that as long as a woman was not a servant, it was acceptable for her to drink alcohol, possibly even to drunkenness.

During the latter of half of the eighteenth century, white Chesapeake men and women increasingly feared insubordinate or vindictive servants and slaves. The focus of much of this fear was alcohol consumption by those slaves. Such apprehensions about lower sort's drinking became possible because nonalcoholic drinks, namely, tea and coffee, were increasingly available. Planters did not give slaves tea and coffee, but the new availability of nonalcoholic drinks changed planters' attitudes toward drinking alcohol. Scholars have not previously noted that it was the introduction of these new drinks that made drinking alcoholic beverages seem like a willful choice and that established the conditions for the temperance movement that developed in the antebellum era.

It is true that there were some fledgling concerns about intoxication before the second half of the eighteenth century. English advice writer Gervase Markham told wives in 1625 that they could cure their husbands' drunkenness by mixing a pound of betany and coleworts together and giving it to them. "As much" of this concoction as "will lie upon a sixpence," Markham promised, "will preserve a man from drunkenness." Most authors did not consider drunkenness to be notable, however, and advice such as Markham's was rare. One prescriptive writer told women that their husbands would be more fond of them when drunk: George Savile assured wives in 1688 that "your husband may love wine more than is convenient . . . in the first place, it will be no new thing if you should have a drunkard for your husband . . . [and] a drunken husband . . . will throw a veil over your mistakes . . . others will like him less, and by that means he may perhaps like you the more."[2]

In America in the seventeenth century, some of the colony's highest elites viewed drunkenness among the lower sort as undesirable. The elites were not concerned about drunkenness per se, but rather about the reputation of the colonies, since they were responsible for enticing settlers. In 1622 the Virginia Company called on Governor Sir Francis Wyatt to "earnestly require the speedy redress" of excessive drinking in the colony, the "infamy" of which "has spread itself to all that have but heard the name of Virginia." The Assembly enacted a few laws against drunkenness that might help to protect their clients' investments in the colony and their reputations. For example, in 1632 the Assembly ruled that "for every offense [of drunkenness] [the offender was] to pay five shillings to the hands of the church wardens." Any defendant who apologized would be excused from punishment.[3]

Typically, though, a member of the elite in the seventeenth or early eighteenth centuries cared little about the drinking of others. Accounts of drunkenness in seventeenth-century Chesapeake court records are unusual because drinking was not remarkable. Even when a case of drunkenness was brought before a court, the alleged defendant was usually excused. For example, in 1654 William Strangnidge was fined fifty pounds of tobacco for being so drunk in court that he committed "abusive, scandalous, rash speech" against Judge Lieutenant Colonel Obedience Robins. When Strangnidge apologized for his behavior the next day, the court remitted his fine and dropped the matter. John Jones drank heavily in court in July of 1660 and began singing and dancing in the courtroom. The court fined him fifty pounds of tobacco, but when he apologized, his punishment was revoked. In March of 1694, Fooke Davis was presented to the court for being drunk on the Sabbath, but when he denied it, the court excused him. Even when Margaret Reine, Captain William Jones's servant, made a false key to the Captain's storehouse and stole a jug of rum from it in 1694, the court voided her punishment "on condition of good behavior." George Stegall was presented to the court in 1736 for drunkenness, but when he admitted the offense, his fine was excused and the case was dismissed.[4]

Judges themselves were frequently drunk in court, a practice the Assembly tried to curb by fining justices who "became so notoriously scandalous upon court days . . . to be so far overtaken in drink." The regulation was ineffective, and no justices were brought up on charges of drunkenness. Ministers were frequently drunk too, which made it difficult for them to condemn their parishioners.[5]

The men presented to Chesapeake courts in the seventeenth and early eighteenth centuries for drunkenness were presented because they were drunk on the Sabbath, drunk in court, drunk in church, or drunk enough to have insulted one of their superiors. Very few women, slaves, or servants were presented for drunkenness during this time because their husbands or masters handled their discipline. When a man was presented to the court for drunkenness, the issue was more the challenge to authority than the intoxication.

The statutory punishment for drunkenness in the seventeenth century was not onerous, and in fact it was the same as the fine for failing to attend church or for swearing: fifty pounds of tobacco or five shillings. To put this fine in perspective, the fine for a woman found guilty of fornication was five hundred pounds of tobacco. The fine for stealing a hog or horse began at one thousand pounds of tobacco, while the fine for killing a cow and eating it was two thousand pounds of tobacco. Drunkenness was a relatively minor infraction in the seventeenth and early eighteenth centuries.[6]

During the seventeenth century and the first half of the eighteenth century, Virginians also accepted drinking and drunkenness among servants while at home. Courts required that masters give their servants and apprentices "sufficient food, [alcoholic] drink, lodging and washing." When colonists noted drunkenness among servants, they usually excused it. William Roberts advertised in the *Maryland Gazette* in 1745 that his servant, John Powell, had not in fact run away, but had "only gone into the country a cider drinking" and was again prepared to repair watches and clocks. Another colonist advertised in 1745 that he had a female servant who was "strong and healthy, can do any household work, and understands weaving," noting that "her principal failing is drunkenness." The seller did not appear to fear that this admission would make her difficult to sell. Landon Carter recorded that his manservant Nassau's drinking was a constant inconvenience, but Carter did nothing about it: "Found Nassau most excessively drunk,"

Carter noted one night, "so that I suffered as much as an old man could suffer" from the lack of the fire that Nassau usually laid in the fireplace. Another time, Carter sighed that "Nassau so constantly drunk," but he concluded only that "there will come a warm day for the punishment of these things." William Byrd laughingly recorded that when his servant, an Indian girl named Jenny, got drunk, she "made us good sport." Byrd also seemed unconcerned about the intoxication of his physician, noting that the man who bled him "was a good natured man but too much addicted to drink." The doctor died an alcohol-related death.[7]

Attitudes began to change by the late eighteenth century. Elites became far more wary of alcohol consumption among the lower sort, particularly the drinking of hard liquor, such as apple brandy. The earliest concerns about alcohol in America arose in the medical community in the 1740s. Physicians, particularly Philadelphian Benjamin Rush, noted a new disease then called the West Indies Dry Gripes. Unbeknownst to Rush, the disease was actually lead poisoning that resulted from the use of lead in the stills that West Indies distillers used to create their rum. Rush's experiences during the Revolutionary War, when he was appointed the Physician-General of the Continental Army, furthered his concerns and led him to publish pamphlets detailing the dangers of liquor. For example, in the 1784 booklet *Inquiry into the Effects of Spirituous Liquors on the Human Body and Mind*, Rush asserted that liquor consumption turned a person into an animal until he resembled "in fetor, a skunk; in filthiness, a hog; in obesity, a he-goat."[8]

Rush's concerns about alcohol went mostly unheeded until the early nineteenth century when Americans began to drink large amounts of corn whiskey. Partly in order to purchase larger farms, Americans had begun heading westward, where corn grew more easily than other crops in the thin, rocky soil of the newly settled areas and could be prevented from spoiling through distilling it as whiskey. The price of whiskey dropped in response to increasing supplies, and Americans could afford to drink more than ever. Most scholars agree that the temperance movement was born of the increasing consumption of whiskey and concurrent increase in drunkenness in the newborn American experiment at the start of the nineteenth century.

A popular story holds that New York citizens reading Rush's pamphlet established the first formal temperance group in the United States in 1808. Other temperance groups devoted to ending the consumption of distilled liquors (fermented drink was viewed with much less concern) blossomed soon after, especially in urban centers of New England.[9]

The developing American middle class, which included those who owned small businesses and manufactories, mostly in New England, found the idea of sober, reliable workers appealing. Until the 1820s employees generally lived with their employers, and employers took responsibility for the welfare of their workers. A new emphasis on family privacy and on the home as a refuge from the workplace led employers to remove workers from their homes. When Presbyterian and Congregationalist ministers began preaching about self-discipline and self-reliance, industriousness and sobriety, the new business class discovered that the ministers' message released them from responsibility for their workers. Employers' wives found the new message exciting as well, because it gave them an important role, instilling self-discipline in the republic. In New England, white middle- and upper-class temperance was seen as a reflection of self-control and respectability, an outward sign of inner grace and a justification for economic well-being. By 1833, over one million Americans had joined more than 6,000 voluntary associations pledged to temperance. By 1855, twelve states in the North had passed laws prohibiting the sale of liquor.[10]

This usual narrative for the temperance movement does not explain when and why *southerners* came to see alcohol as a problem. Southerners were generally more hesitant about temperance until the 1850s because the temperance movement was strongly associated with abolitionism. Benjamin Rush was also president of the nation's first antislavery society. Moreover, the South was so rural that it was difficult for southerners to get or obtain drink other than alcohol. The region did, however, develop more than three hundred temperance societies, often in the most urban sections after 1826.[11]

Scholars have offered a variety of thoughtful and persuasive explanations for the rise of the temperance movement in the North. No one has considered the impact of the availability of nonalcoholic drink, however. In the Chesapeake, the slow penetration of tea and coffee into the marketplace helped create the conditions that allowed temperance to flourish later. Upper-sort Chesapeake men and women began to voice limited concerns about alcohol, especially distilled liquor, beginning in the 1760s, in part because tea and coffee were becoming more available in the region.

Coffee and tea became fashionable among the wealthy in the Chesapeake because those drinks were fashionable in Europe. Court aristocracy had begun drinking the new luxury beverages of chocolate, coffee, and tea starting in Venice, Marseilles, London, and Amsterdam in the early seventeenth century. Drinking coffee from porcelain cups at court was thought to demonstrate personal elegance. Soon doctors and prescriptive writers began to ascribe virtues to coffee that they had previously assigned to alcohol. Coffee appealed to the new age of enlightenment for the way it supported sober, rational thought and discourse. European merchants seized on the new beverages, and by the late seventeenth century, advertisements proclaimed that coffee purified the blood, balanced the humors, fortified the liver, promoted "the disappearance of the spleen," and offered "relief against violent headaches and vertigo." Coffee drinking spread to the middle class through coffeehouses, and by 1700, London alone boasted 3,000 coffeehouses. By 1750, middle-class households in Europe began to brew coffee.[12]

Following quickly after coffee came tea, and by the middle of the eighteenth century, the English were drinking mostly tea. English leaders preferred tea for their nation because English colonies grew tea. English men and women seem to have preferred tea to coffee as well, likely because tea was cheaper because it took fewer tea leaves than coffee beans to make each cup. Tea could be easily re-steeped to make another cup. Sir Kenelme Digbie advised that "the water is to remain upon it [the tea leaves] no longer than whiles you can say the Misere Psalm very leisurely," so that the tea leaves could be used for another cup. The cost of tea continued to drop throughout the eighteenth and nineteenth centuries.[13]

But why did elite Virginians want to emulate Englishmen when they had been out of step with England and English drinking habits for so long? The colonists had always wanted to be English people in America, but the tobacco monoculture and the resulting lack of marketplaces had held them back. Tea drinking was something colonists could take part in. Chesapeake men and women who had the time and money to drink tea could participate in genteel English life and experience "refinement."

The codes of behavior that defined the new refinement had begun with Renaissance European courtiers. Courtiers learned to restrain their bodies, cultivate feelings, and give dramatic recognition to rank; and they accompanied these behaviors with new, sumptuous households containing carpets, clocks, mirrors, individual cutlery, rich fabrics, and newly invented musical instruments. The new manners and accoutrements finally spread to the elite in the American colonies after 1690. Although colonists could not obtain or afford all the trappings of refinement, tea drinking allowed them to discuss the new ideas and novels from Europe while they displayed their new clothes and matching tea cups from England. Tea drinking legitimized the claim of being refined English gentry in Virginia.[14]

Coffee and tea entered the Chesapeake in the mid-eighteenth century through the most fashionable homes and taverns. In 1754 Jonathon Greenhow purchased the license to Marot's Ordinary in Williamsburg and changed it from a tavern that sold alcoholic drinks to a coffee house that was popular among Virginia's wealthy. Other taverns that catered to elite clientele added tea and coffee to their menus. Anne Pattison also added tea and coffee equipment to her fashionable Williamsburg tavern, and large planter Robert Wormeley Carter recorded paying for tea at Mrs. Richardson's tavern.[15]

By the end of the century, tea was available in most taverns in the Chesapeake, and travelers recorded drinking tea as part of a tavern breakfast in the region. Ferdinand Bayard explained that by 1791 the "usual traveler's breakfast" at a tavern was "ham, broiled chickens with a cream sauce, slices of bread spread with butter, tea, and coffee." Traveler Robert Honeyman found tea and coffee in taverns at all times of the day in 1775. His diary reveals that he ordered tea or coffee any time a traveler of an earlier era would have had an alcoholic beverage. Likewise, Alexander Macaulay recorded drinking tea at the end of a long traveling day, a time when previous travelers would have drunk rum punch. "Arrived a little in the night safe at Byrds Tavern," he wrote in his journal; "we had a dish of tea." Just as making alcoholic beverages had once been women's work, roasting coffee beans and preparing coffee and tea were also women's work. Harry Toulmin recorded during his travels throughout Virginia that "I have never yet seen a gentleman make tea. If there be no female of the party, a servant waiting at the table makes and pours out the tea for everyone."[16]

Sales of tea and coffee in taverns mirrored changes in elite households. Because tea production was localized in the British colonies, its price became low enough for the upper sort to drink it daily by the mid-eighteenth century. Tea and coffee became the drinks of choice for social gatherings among those who could afford them. Luicinda Orr, a young lady, noted in her journal in 1782 that "this evening Colonel Ball insisted on our drinking tea with him: we did, and I was much pleased with my visit." Her brief journal makes clear that all of her socializing involved tea or coffee. Frances Baylor Hill's short adolescent diary suggests that tea and alcohol were already being distinguished along racial lines. "Tom Ta-or would insist on our going to Beargarden to drink tea with him," she recorded of her visit to Pleasant Hill plantation. "The gentlemen made an old negro woman drunk and then she turn'd cut and shew'd proper capers dancing and maneuvering about," she noted disapprovingly.[17]

Wealthy plantation mistress Ann Manigault confirmed that tea was the main focus of social gatherings of elite women and couples. "Drank tea at Mrs. Wragg's," she recorded in her diary on June 15, 1755. "My son and his wife drank tea here and rode out," she wrote a few days later. Margaret Bayard Smith recorded endless rounds of teas in early Washington, some with ladies only and some with men as well. Eleanor Parke Custis Lewis, George Washington's adopted daughter, likewise spent her social engagements drinking tea instead of alcohol. She explained to her friend Elizabeth Bordley that she had attended two small tea par-

ties in Alexandria, Virginia: "Saturday dined out and drank tea also . . . Sunday —
ditto." Susanna Knox traveled to Winchester, Virginia, to care for relatives with
smallpox. She found time every day to visit someone for tea or to have women
over for tea. One day she drank tea "with a large party of ladies" at Miss Baldwins,
and the next day she drank tea at Mrs. Tidbald's.[18]

Tea or coffee replaced alcohol as the drinks of choice for all times of the day
among the elite. Large planter William Byrd's diary reveals that by the mid-
eighteenth century the Chesapeake's wealthiest residents had traded their morning cider or ale for tea or coffee, and that the same was true for social occasions.
"I rose about 6," he recorded on December 6, 1739. "I prayed and had coffee."
"Had tea at Colonel Grymes's. . . . After dinner we visited Colonel Lightfoot
where we drank tea," Byrd noted on another day. John Harrower, who was a tutor
and indentured servant in Virginia in the 1770s, wrote to his wife that he ate at the
planter's table and had coffee with breakfast. Olney Winsor, while conducting
business in Virginia, stayed in a boarding house in Alexandria where "I breakfast
on coffee — dine on meats — sup on tea or milk." Plantation mistress Martha Ogle
Forman and her husband thought that a family friend should not be without coffee when there was a death in the family and so "purchased for Mrs. Thomas a
shroud and cap and other things necessary for [the] funeral with 6 pounds of coffee and 6 pounds of sugar." Moreau de St. Mary declared during his travels to
America that "tea is always served at the first meal; and it is this leaf, passionately
loved by the Americans, which constitutes the entire third meal." Finally, Sarah
Cary at Mount Pleasant plantation wrote to a friend that she had been so "low
spirited, alone, and unwell" that she did not drink her tea at breakfast.[19]

Tea became so ingrained in the culture that during the American Revolution,
some Loyalist families invited soliders to tea. "Capt. Lipscombe drank tea with
us," Sarah Wister recorded in 1777 while the soldiers were stationed in the area.
Sarah Frost had tea even when fleeing the American army: "I left Lloyd's Neck
with my family and went on board the *Two Sisters*, commanded by Capt. Brown,
for a voyage to Nova Scotia with the rest of the Loyalist sufferers. This evening the
captain drank tea with us."[20]

Probate records and diaries reveal that coffee-making and tea-making utensils
began to spread among the middling classes in the Chesapeake during the latter
half of the eighteenth century and into the early nineteenth century. For example,
the 1764 probate record of small-planter Joseph Bradford states that he owned a
table, four chairs, an iron pot, and a coffee pot. Similarly, Thomas Griffin Peachy
purchased a coffee pot for four shillings in 1796. Researchers have found that
equipment for making and drinking tea appeared in a wide variety of Chesapeake

households by the time of the American Revolution. The percentage of houses with tea equipment in the Chesapeake was slightly behind that of Massachusetts, but the Chesapeake was catching up quickly. A traveler who toured the new federal buildings in Washington, D.C., recorded that "I went into the hut of one of these workers. I found his wife there dressed very neatly, good utensils for cooking, and all the service for tea in porcelain from China." He found equipment for tea in some slave quarters as well: "We entered one of the huts of the blacks, for one cannot call them by the name of houses," he noted, "they are more miserable than the most miserable of the cottages of our peasants. The husband [and] the wife sleep on a mean pallet, the children on the ground; a very bad fireplace, some utensils for cooking, but in the middle of this poverty some cups and a teapot."[21]

Once the upper and middling sort had something nonalcoholic to drink, they began expressing concerns about drunkenness. Almanacs began to urge prudence in drink. Benjamin Franklin cautioned in his popular almanac: "He that spills the rum, loses that only; / He that drinks it, often loses both that and himself." Theophilus Wreg warned in *The Virginia Almanac* for 1766:

> Cold winter's pinching weather now begins,
> Keep up good rousing fires to warm your shins;
> Eat well, drink well, but nothing out of measure,
> But earn it first, and 'tis enjoyed with pleasure.

The author of *The American Calendar* for 1771 concluded: "Yet faith! my friend! in sober sadness, / Drinking's a stupid kind of madness," and:

> Bottles and girls are not the things
> From which perpetual pleasure springs;
> Hard drinking is a heavy duty,
> And surfeits may be got by beauty.

The widely read *Virginia Gazette* likewise began cautioning against over-imbibing, asserting that the sober man develops "sprightliness of his mind, warmer and better founded than any derived from wine. . . . [He] goes to bed satisfied . . . whereas he who drinks falls asleep without knowing it, is uneasy when he wakes, and vexed at being mad yesterday, makes himself mad today, that he may forget it. . . . [A] man drinks a glass or two at his meals with a proper relish . . . carried farther, the blessing is lost." One Dr. Allen advertised that he had a tonic to counteract "the drinking of spirits and strong waters [that] is become very common amongst the people of inferior rank" and caused "the destruction of their healths, enervating them, and rendering them unfit for useful labor, intoxicating them, and debauching their morals, and leading them into all manner of vices and wickedness."[22]

Households began to buy decorative pictures that warned of the dangers of drunkenness. Cheap prints flooded the Chesapeake in the mid-eighteenth century as competition among London print sellers reduced prices dramatically. Some of the most popular prints in England and the Chesapeake were the "prodigal son series," a group of six inexpensive prints that illustrated how adolescent men ought and ought not to behave through the story of the danger of leaving home. One visitor touring America recorded in Hanover, Pennsylvania, not far from the Maryland border, that in most houses "the decorations on the wall," were usually "representing the story of the prodigal son. From this story the fathers give to their sons all their moral instruction." The way the prints dealt with alcohol changed over the course of the eighteenth century. In the prints from the 1750s, the son who left home and fell into sin and poverty is welcomed upon his return with a large feast, featuring copious alcoholic drinks. By the 1790s, however, the prints no longer showed alcoholic beverages in the scenes of reunion and proper family life, but only in the scenes of the ruin of the prodigal son.[23]

Once Chesapeake elites had nonalcoholic drinks available and began to learn about the problems blamed on alcohol, they began to desire sobriety in their servants. During the latter half of the eighteenth century, elites began to stipulate that their employees be "sober," a word that began to mean not only serious in demeanor but also free from intoxication. For instance, Francis Jerdone, a Virginia merchant, asked his father in England to engage a house carpenter for him "that is a good tradesman, and withal sober and industrious." Joshua Johnson complimented himself for having engaged two "sober, honest, industrious men" for his mercantile firm. Thomas Hall advertised for "a sober person, of good morals" to be a Prince George County schoolmaster. Philip Carbery announced that he would rent out a slave who was "by trade a barber, and remarkable for his carefulness, sobriety and honesty." Another "sober, very honest," slave was for sale as a house servant. A Virginia tavernkeeper advertised for "a sober, discreet, honest man" to be a drawer of drink. Another Virginian advertised that he "wanted to hire, a steady negro man . . . who can be well recommended for his . . . sobriety." Virginia planter William Fitzhugh entreated his friends to procure "an able, learned, serious, and sober minister." "I hope you remember," he wrote, "to persuade a minister in, if you can with a sober, serious learned one." Heavy drinking was not uncommon among ministers at the time as Englishwoman Charlotte Browne noted of her Atlantic crossing to America in 1754: "Sunday but held no prayers, our parson being indisposed by drinking too much grog the night before."[24]

The circumstance that allowed drunkenness to become a matter of class was the fact that nonalcoholic drinks were very expensive. Tea and coffee required pricey ingredients, cooking utensils, and plateware. Even middling-level families could find it challenging to get tea leaves and coffee beans. Francis Taylor noted that in 1792 when his father wanted coffee, he had to wait until a relative went to Fredericksburg and purchased a half pound of coffee for him. When Robert Wormeley Carter wanted tea, he had to order it from London merchants. Making coffee then required time and attention that many people could not afford. One recipe to make coffee advised that "first you brown and grind the beans, mix 6 tablespoons of ground coffee with an egg white, stir in cold water to moisten, let stand 1/2 hour, then pour on the mixture 12 teacups of boiling water, set on stove to simmer, then strain off."[25]

Not surprisingly, middling families often did without the luxurious drinks. Devereux Jarrat recalled that his middling parents "always had plenty of plain food — wholesome and good, suitable to their humble station. . . . Our food was altogether the produce of the farm or plantation, except a little sugar, which was rarely used. . . . We made no use of tea or coffee for breakfast, or at any other time; nor did I know a single family that made any use of them. . . . I suppose the richer sort might make use of those and other luxuries, but to such people I had no access." While middling households rarely could afford the new caffeinated beverages, servants and slaves almost never could.[26]

The mounting association of liquor consumption with foul behavior developed at the same time that planters and merchants were growing increasingly apprehensive about slave uprisings and retaliation, in the late eighteenth century. Chesapeake newspapers ran stories of servants and slaves who harmed their masters or the public while drunk. A December 1767 issue of the *Virginia Gazette* included a story from New York, in which a group of twenty black and white servants and slaves were said to have gathered at a "poor white man's house" with two pigs for barbecuing and two gallons of wine for a "junketing frolic." According to the story, the group did not steal or damage anything, but they were whipped at the public post, and the newspaper castigated the white man for "ruining servants." The fact that the *Virginia Gazette* printed the story illuminates the growing fear of Chesapeake planters that blacks and poor whites might unite, particularly if they were allowed to drink alcohol and socialize together unsupervised. This type of story became common following the slave revolts at Stono, South Carolina, in 1739 and in Jamaica in 1776, and the series of black revolts in the British West Indies in the 1770s. The slave conspiracies in New York City in 1741, at Point Coupee, Louisiana, in 1795; in Richmond, Virginia, in 1800; and in Southampton County, Virginia, in 1831, stoked white fears.[27]

Whites increasingly feared that slaves who drank would become uncontrollable. Portraying slaves and servants as without discipline justified their need for masters, at least in the eyes of the upper sort. Thomas Jones, a Welsh servant who had run away, was advertised by his master as "backwards when he is in liquor." A runaway convict servant man was "very quarrelsome when drunk," and runaway James Jordan "loves liquor." Mulatto slave Jack "is much given to liquor, he is when drunk very talkative and quarrelsome"; while runaway English servant William Springer, according to his master, even "broke open my cellar and stole a large quantity of rum."[28]

Planters often stressed, in late-eighteenth-century advertisements, that slave and servant runaways were "addicted" to drink, meaning prone to drunkenness. This was a way of arguing that the planters were not bad masters. Masters looking for runaway slaves and servants noted that alcohol made their servants "impudent," "artful," "ill behaved," or "ungrateful." Neil Buchanan advertised that his runaway slave "talks much, especially when in liquor, to which he is pretty much addicted." Mulatto servant Jack "is much addicted to drinking, which makes him forward"; while Daniel, a mulatto slave, "loves drink so well, if he can get it he will be drunk." Another mulatto slave, Dick, was also "much addicted to liquor," and his master warned ship captains against assisting "so notorious a villain; for when he is drunk, he will steal anything." Thomas Jefferson pronounced that a runaway slave of his was "greatly addicted to drink, and when drunk is insular and disorderly, in his conversation he swears much, and in his behavior is artful."[29]

Masters believed that they provided all the alcohol that was good for their servants and slaves, and that anything else represented a challenge to their authority. Masters wrote in their diaries and letters that they routinely offered slaves and servants drams of cider and rum during harvest, at holidays, and after particularly difficult work. For instance, Landon Carter recorded that "I gave as much toddy, beer, and cider [at a gathering] as could be drunk and I believe the quantity used will show it to be a full plenty." William Byrd congratulated himself in his diary for providing his slaves "some rum and cider to be merry with." Too much drink simply made servants disrespectful. William Gregory angrily noted that his slave, Peter, was "much delighted when he gets any strong drink, which he is remarkably fond of, and then very talkative and impudent." Francis Hagure argued that his servant man "is a lover of strong drink, and [the] very first to take to when opportunity offers, and is then very ill behaved." The mulatto slave named Dick ref-

erenced above was "extremely fond of liquor, and when drunk very impudent."
Miriam Richardson's slave was similarly "very fond of strong liquor, and very talkative and saucy when drunk." Harry, another slave, was "fond of liquor, and when
drunk very impertinent." Robert Rutherford denounced an "ungrateful villain,"
his Irish servant, who "is fond of spirituous liquors, and a remarkable reprobate
when drunk."[30]

Increasingly, masters punished servants and slaves who drank outside the
boundaries prescribed by their class. While William Byrd could get drunk at the
Governor's Palace, noting in his diary that "after dinner waited on the Governor,
stayed about an hour and walked home, a little giddy," he could also beat a slave
for similar behavior. "My man Tom got drunk and did not what I bade him; so I
beat him," Byrd recorded.[31]

To slaves, it was white men and women who could not control themselves
when drinking. James Curry recalled that his overseer, "although he seldom
whipped his slaves cruelly, at times, when he began to whip a slave, it seemed as
though he never knew when to stop. He usually was drunk as often as once a
week, and then, if anything occurred to enrage him, there was no limit to his
fury." Curry concluded that his "master and mistress were both drunkards." Lewis
Clarke, another enslaved man, said of his plantation mistress that "she used to sit
over her toddy, trying to invent some new way to punish 'em. Master was a little
too fond of grog; she used to keep it locked up from him; and he had to coax her
to get any. Sometimes, when he came home, she would whine and groan about
what a hard time she had of it; and tell how the slaves acted so unruly she couldn't
manage 'em. 'Well, give me a dram,' he'd say, 'and I'll beat 'em for you.' "[32]

James Fisher recalled that his owner "was a drinking man, and when he was
in liquor, sometimes abused me cruelly. When he came sober, he never seemed
to know he had done anything wrong." Former slave James Madison declared that
his owner was so incompetent in the face of liquor that when he and an assistant
captured Madison after an attempted flight, both "were so overjoyed at the recap-
ture of their victim, that they stopped at almost every public house on their way,
to drink and boast about what they would do with him when they reached home.
After they got about half-drunk they bestowed very little attention on their mana-
cled victim, whom they supposed was perfectly safe at the hind part of the buggy."
Madison escaped again. James Smith revealed a similar experience after he ran
away: when his owner's slave-catchers found him, they indulged in celebratory
whiskey. "The whole crowd drank whiskey so freely that night that they became
stupid and careless about Smith, after they supposed that they had got him
drunk. . . . Several of them said that he was so drunk that he would not be able to

stir before the next morning, so they retired and left him lying on the kitchen floor. . . . About one hour afterwards when he supposed that all were asleep he bid final adieu to the abodes of slavery" and escaped.[33]

Levi Douglass was also captured and imprisoned for running away. At the jail, according to Douglass, white men and women threw coins at him when he danced. Douglass used the money to buy whiskey for his jailer and escaped from jail "when his keeper was in a state of complete intoxication." Sella Martin was sold to a slave trader who "had a secretary who was addicted to the excessive use of strong drink, the love of which had grown upon him." The secretary was dismissed, and Martin was sold to another trader who also had a drinking problem. "I was often compelled to get aid to carry him to his room; but, under the stupefying effects of drink, he soon became a victim to gamblers," Martin recalled. Finally, plantation mistress Martha Forman recorded that after a slave catcher caught her runaway slave and "secured him with handcuffs and tied him with a rope" at Pearce's tavern, the slave "slipped his handcuffs, untied his ropes and got out of the room," leaving the slave catcher drunk and asleep. According to many former slaves, it was white men and women who could not control their drinking.[34]

In one sense the slaves were right. The Chesapeake's male elite delighted in bragging about their own drinking exploits. This was a new practice. Earlier generations had not boasted of how much they drank, because alcohol was something they drank throughout the day rather than in particular episodes. Now, planters celebrated their own bouts of drunkenness even as they sought to control slaves impertinences. James Kirke Paulding told Morris Smith Miller that when he was in Washington, he would "have some potential bouts at the mint juleps," and that he would share "a secret by which you may get safely home after drinking six bottles. It is by just putting your feet on the edge of the table, by which means the wine is prevented from descending into the legs, thereby making them as drunk as nine pins. I have tried this method several times and do assure you, that . . . you may drink up to the chin and afterwards walk home as steady as a church steeple. This is a most invaluable secret, and I advise you strenuously to put it in practice." William Fitzhugh wrote "I am hurried away and hard in drinking with two masters."[35]

The Tuesday Club, a mid-eighteenth-century social club founded by Dr. Alexander Hamilton in Annapolis, celebrated its members' abilities to imbibe. It and similar clubs were founded for the purpose of allowing elite white men to gather and drink. The club drank toasts on any pretext at each meeting, including toasts to the long life of the club, to the ladies, to any military successes, and to the health of various royal personages. The club drew up numerous tongue-in-cheek rules that illustrate the emphasis on drinking: if a member did not attend a club meeting, then he owed a half-crown fine to the club to be spent on "rack or other liquors." The club placed one member on trial when he did not bring the club some of the beer he had received from England. Dr. Hamilton paid one of many tributes to alcohol and drinking on the club's fourth anniversary, declaring that here are the "punch bowls always replete with fragrant and nectarous liquor, for this cordial juice . . . heightens the spirits, enlivens the wit, and will conduce not only to make me a more fluent orator, but you more jolly and benevolent *long-standing members.*"[36]

In another example, the *Virginia Gazette* published an account of an impromptu celebration at a tavern after an Englishman who was known for his "attachment to the interests of America" was elected to Parliament. The writer emphasized that the planters drank at the tavern "to testify their joy on the occasion." Newspapers published other tributes to elite male drinking in poems, for example:

> With you, kind sir, to drink the generous glass,
> and o'er a social bowl our evenings pass,
> To banish all the sullen gloom of night,
> In cheerful, innocent, yet gay delight,
> while truth and friendship animates the soul.

The poem concluded that "'tis wit and friendship form the joys of wine." In another of many such paeans, a man thanked his patron for time the patron had spent with him: "In social mirth some spend the hours, / And quaff rich wine in fragrant bowers." Benjamin Franklin paid a humorous tribute to alcohol as well:

> To die's to cease to be, it seems,
> So learned Seneca did think;
> But we've Philosophers of modern Date
> Who say 'tis Death to cease to Drink.[37]

Increasingly, these men who believed that "'tis death to cease to drink" expected colonists below planter status to remain sober. This change of expectations was possible because of the spread of nonalcoholic drinks into the region over the second half of the eighteenth century, and it established the conditions for the temperance movement that would flourish in the nineteenth century. The spread of nonalcoholic drinks, then, made drinking alcohol appear a conscious decision, one that planters came to view as a challenge to their authority when it was made by anyone they expected to control. It would not take much to convince planters that those beneath them should drink no alcohol at all.

Conclusion

Until the late eighteenth century in the Chesapeake, alcoholic beverages were considered a necessary part of daily life for health and socializing. Men valued women who could make alcoholic beverages and depended on this aspect of women's cookery for survival and pleasure. By 1782, however, Luicinda Orr, an adolescent in Virginia, could record in her diary that she was "in a peck of troubles for fear" that the male guests staying at her plantation home "should return drunk." When the gentlemen arrived and were, indeed, "tipsy," she and her friend ran off to avoid them. Becoming tipsy had become something to disapprove of, and it was no longer women's work to make the drinks.[1]

The eighteenth century saw other changes too, many sparked by developments in technology. The invention of the notion of "science" and technological innovations such as hops, the alembic still, improved cider presses, and the Hewes crab apple led men in Europe and America to claim that alcoholic beverage production was chemistry, not cookery, and belonged to men's domain. Although the Chesapeake lagged one hundred years behind the rest of the Atlantic world in this transition, it did slowly catch up, particularly when the Continental Army of the American Revolution helped spread the new thinking.

Women did not resist this change; given that they had plenty to do and that they had not earned money from making alcohol because of the Chesapeake's rural character, there was no reason for them to contest men's assumption of women's work. Women did continue to manage most of the taverns in the region, despite the fact that the tavern licenses were in male names. Tavernkeeping was something that women could do as they managed their households.

Men adapted alcoholic beverage technology to different ends. Large planters used technology to make ever larger amounts of alcoholic beverages for sale, although many found profits elusive, particularly after small planters stopped buying from them in the latter half of the eighteenth century. Small planters had once depended on large planters for drinks when women's household production was not sufficient. In the mid-eighteenth century, though, small-planter households used technology to create a drink trade with people of their own class, severing their dependence on large planters and their small connection with the Atlantic world in order to exchange and sell alcohol with people of their own kind.

Finally, drinking alcohol at all became a choice. Until the late eighteenth century there was, simply put, nothing nonalcoholic to drink in the Chesapeake. The spread of tea and coffee into the region, and their increasing affordability, meant that as the century wore on, drinking alcohol began to appear to be a personal choice. Upper-sort colonists ignored the fact that coffee and tea cost more than alcohol and chose to see alcoholic beverage consumption as a personal preference or bad decision. Slave owners in particular began to punish slaves and servants for drinking alcohol. The region would not join the temperance movement until the 1830s, but late-eighteenth-century colonists created the preconditions that allowed the temperance movement to flourish later.

Given that for centuries women had made alcohol, it is perhaps ironic that women led both the temperance and prohibition movements in America in the early nineteenth and early twentieth centuries. In the temperance movement, especially in New England where the economic and social changes wrought by industrialization were greatest, white middle-class women pressed men to "take the pledge" and forsake alcohol. Women thought that they were helping to make the American republic successful by encouraging men to restrict their drinking. More or less a century later, women amplified their alcohol-limiting activities and combined the fight for women's suffrage and for more limited drinking with the prohibition movement, inherently arguing that women's voting and women's vigilance against alcohol would keep the republic pure. The suffragists and prohibitionists were successful. Women gained the right to vote with the ratification of the Nineteenth Amendment in 1920. Following many state laws and constitutional amendments, Congress approved the prohibition of the manufacture, sale, or transportation of intoxicating liquors in the United States and its territories in January of 1919. While drinking itself was not illegal, for a time making, selling, or transporting alcohol was (Congress repealed prohibition in 1933).

While it is possible to speculate that women fought to control alcohol consumption in the nineteenth and twentieth centuries because they had lost control of alcohol production and distribution in the eighteenth century, there is no evidence of such a relationship. Women were not disgruntled to have "lost" alcohol production; being able to purchase alcoholic drinks, tea, and coffee made their lives much easier. If men wanted to bend over boiling kettles and steaming stills, so much the better. In fact, by the early twentieth century, most American women and men had forgotten that making alcohol was once part of women's cookery. It is well worth remembering, though, that making alcohol used to be women's work.

A Few Recipes

GINGER BEER

"Ginger Beer by a receipt from Mrs. Colonel Chambers" (1819). "3 Gallons of water, 3 large tablespoonfuls of ginger, 2 lb. of brown sugar, one teacup full of good yeast. Let it stand 24 hours, when strain it through a thick bag, and bottle, upon 24 hours you may use it." From Martha Ogle Forman, *Plantation Life at Rose Hill: The Diaries of Martha Ogle Forman 1814–1845*, ed. W. Emerson Wilson, 3 Sept. 1819 (Wilmington: Historical Society of Delaware, 1976).

PERSIMMON BEER

"Mr. Jefferson's receipt to Make Persimmon Beer" (1822). "Gather the persimmons perfectly ripe and free from any roughness, work them into large loaves, with bran enough to make them consistent, bake them so thoroughly that the cake may be brown and dry throughout, but not burnt, they are then fit for use; but if you kept them any time you must dry them frequently in an oven moderately warm, of these loaves broken into a coarse powder, take eight bushels pour on them forty gallons cold water and after two or three days draw it off, boil it as other beer, hop it, and it makes a very strong beer; by putting thirty gallons of water in the same powder and letting it stand two or three days longer you may have a very fine small beer." From Peter Cottom, *Virginia and North Carolina Almanak for 1822*, MS, Virginia Historical Society, 1821, 36.

CIDER

"Pull your fruit before it is too ripe, and let it lie but one or two days, to have one good sweat; your apples must be pippins, pearmains or harveys (if you mix winter and summer fruit together, it is never good) grind your apples, and press it; when your fruit is all pressed, put it immediately into a hogshead, where it may have some room to work, but no vent, but a little hole between the hoops, but close bunged; put three or four pounds of raisins into a hogshead, and two pounds of sugar, it will make it work better; often racking it off is the best way to fine it, and always rack it into small vessels, keeping them close bunged, and only a small vent hole; if it should work after racking, put into your vessel some raisins for it to feed on; and bottle it in March." From Eliza Smith, *The Compleat Housewife* (1758; London: Studio Editions Ltd, 1994), 251.

STRONG BEER

"Take two bushels of malt, and half a bushel of wheat, just cracked in the mill, and some of the flour fitted out of it; when your water is scalding hot, put it into your mashing vat; there let it stand till you can see your face in it; then put your malt upon it; then put your wheat upon that, and do not stir it; let it stand two hours and a half; then let it run into a tub that has two pounds of hops in it, and a handful of rosemary-flowers; when it is all run, put it in your copper, and boil it two hours; then strain it off, setting it a cooling . . . clear it very well before you put it a working; put a little yeast to it, when the yeast begins to fall, put it into your vessel; and when it has done working in the vessel, put in a pint of whole wheat, and six eggs; then stop it up; let it stand a year, and then bottle it." From Smith, *Compleat Housewife*, 254.

SPRUCE BEER

"Take about half a pound of spruce or common pine tops, half a pound of china root, half a pound of sassafras and one quart of indian corn. Put all these ingredients into seven gallons of water and let it boil away to five gallons or till the corn begins to crack open. Take it off the fire and let it stand till 'tis cold, then put it into a cask with about a pint of yeast or grounds of beer and three pints of molasses, and when it begins to work bottle it." From Harriott Pinckney Horry, *A Colonial Plantation Cookbook: The Receipt Book of Harriott Pinckney Horry*, ed. Richard J. Hooker (1770; Columbia: University of South Carolina Press, 1984).

GINGER BEER

"Take of water 6 gallons; brown sugar 5½ pounds, brandy 2 quarts; lemon peels 1 dozen; a race of ginger, 3 ounce-and one pint of yeast. The yeast to be put in the keg first, and the other ingredients to be boiled all together, and suffered to stand until milk-warm; then to be poured on the yeast and left to ferment for 24 hours. Then stop the bung hole and let settle a day or two before you bottle it off." From *The American Farmer*, ed. John S. Skinner, 25 June 1819.

QUINCE WINE

"Take quinces wipe them, and beat and strain them, then put to the meat that you have strained the liquor out of, 3 pints of water if you have had 3 liquor first from it, put 2 pounds of sugar to 3 quarts of liquor, and boil it in the last liquor keeping it [?] when it is cold put the liquors together, with as much yeast as to work it until it is clear, then draw it off into bottles, if you have but little, otherwise into a vessel fit for that quantity you have let stand 14 days or 3 weeks before you draw it off. If it be not sweet enough put more sugar. Remember you may drink it in 6 weeks but will keep a year." From Hannah Huthwaite, *Recipe Book, ca.* 1720, MS, Winterthur Library.

BIRCH WINE

"When the sap of the birch-tree will run, cut a large notch in the bark of the trunk of the tree. . . . You may expect about a gallon of liquor from each of them, which must be ordered in the following manner. Take five gallons of the liquor, to which put five pounds of powder-sugar, and two pounds of raisins of the sunstoned [sundried grapes, not currants]; to this, put the peel of one large lemon, and about forty large fresh cloves: boil all these together, taking off the scum carefully as it rises; then pour it off into some vessel to cool, and as soon as it is cool enough to put yeast to it, work it as you would do ale for two days, and then tun it." From Richard Bradley, *The Country Housewife and Lady's Director*, ed. Caroline Davidson (1727; London: Prospect Books, 1980), 39.

GOOSEBERRY WINE

"To every pound of gooseberry when picked and bruised put one quart of water let it stand 3 days stirring it twice a day. To every gallon of liquors when strained add three pound of loaf sugar let it stand six months in the barrel before you bottle it off. To every 20 quarts of wine put one quart of brandy and a little [sugar?]. The wine should be made just before the fruit begins to ripen." From Philip Evans, *Recipe Book* 1793, ms., Winterthur Library.

YEAST

"Make a strong hemp-tea, thicken it (as you would with potatoes) with flour, and when it ferments, work in as much meal as you can, then make it in cakes as large in circumference and thick as a cracker, and dissolve about a third of one in water to make quart of flour in bread, the cakes must be dried in the shade. It will keep for six months." Ann S. Tazwell to her sister, 28 Aug. 1820, *Tazewell Family papers* 1782–1832, photostat, Rockefeller Library.

Notes

ABBREVIATIONS

APS	American Philosophical Society
CWF	Colonial Williamsburg Foundation
HSP	Historical Society of Pennsylvania
IEAHC	Institute of Early American History and Culture
LC	Library of Congress
LCP	Library Company of Philadelphia
LVA	Library of Virginia
OIEAHC	Omohundro Institute of Early American History and Culutre
VHS	Virginia Historical Society
VMHB	*Virginia Magazine of History and Biography*
WMQ	*The William and Mary Quarterly*

INTRODUCTION

1. "Jack Daniel's history" at www.jackdaniels.co.UK/oldno7 (viewed 6 Feb. 2009).

2. David W. Conroy, *In Public Houses: Drink and the Revolution of Authority in Colonial Massachusetts* (Chapel Hill: Published for the Omohundro Institute of Early American History and Culture [OIEAHC] by the University of North Carolina Press, 1995); Peter Thompson, *Rum Punch and Revolution: Taverngoing and Public Life in Eighteenth-Century Philadelphia* (Philadelphia: University of Pennsylvania Press, 1999); W. J. Rorabaugh, *The Alcoholic Republic: An American Tradition* (Oxford: Oxford University Press, 1979).

3. Judith M. Bennett found that brewing gave women in medieval England a source of income and power in *Ale, Beer, and Brewsters in England: Women's Work in a Changing World* (Oxford: Oxford University Press, 1999). Jane E. Mangan also determined that making and selling alcoholic drinks gave women authority in *Trading Roles: Gender, Ethnicity, and the Urban Economy in Colonial Potosí* (Durham, NC: Duke University Press, 2005). See also Richard Unger, *A History of Brewing in Holland 900–1900: Economy, Technology and the State* (Leiden: Brill, 2001).

4. In contrast to male magistrates' efforts to keep tavernkeeping in the hands of middling women in the Chesapeake, David Conroy found that in seventeenth-

century Massachusetts magistrates awarded tavern licenses to poor women to keep them off the poor (tax) rolls. My findings differ as well from Linda Sturtz, who has posited that eighteenth-century Virginia women from all classes and situations obtained tavern licenses, which they used to run taverns as a source of "implicit power" over men. Sharon Salinger likewise has emphasized the variety of people who obtained tavern licenses in early America. In fact, Salinger declared that, across the colonies, "the most salient characteristic of licensing practices in the eighteenth century was its arbitrariness," and called tavern licensing a "random pattern of administration." Kathleen Brown, on the other hand, has maintained that in eighteenth-century Virginia, women possessed an "increasingly marginal position in the male world of the tavern." See Conroy, *In Public Houses*; Linda Sturtz, *Within Her Power: Propertied Women in Colonial Virginia* (New York: Routledge, 2002); Sharon V. Salinger, *Taverns and Drinking in Early America* (Baltimore: Johns Hopkins University Press, 2002), 180; Kathleen M. Brown, *Good Wives, Nasty Wenches, and Anxious Patriarchs: Gender, Race, and Power in Colonial Virginia* (Chapel Hill: Published for the Institute for Early American History and Culture [IEAHC] by the University of North Carolina Press, 1996), 284.

5. I argue that women did not lose out when men assumed control of alcoholic beverage production in the latter half of the eighteenth century. Most scholars of colonial women have agreed, however, on a declension model stating that after a moment of potential gender equality in the seventeenth century, colonial women lost their public roles and authority during the eighteenth century. Cornelia Dayton found that women lost access to courts in eighteenth-century Connecticut. Deborah Rosen argued that laws of coverture and inheritance "worked to marginalize women from the growing [eighteenth-century] economy." Jeanne Boydson reasoned that the value placed on women's domestic labor in the seventeenth century declined during the eighteenth century. Accordingly, while "the colonial goodwife [was] valued for her contribution to household prosperity," by the mid-eighteenth century, "the denigration of women's household labor was becoming an established cultural practice." Cynthia Kierner concluded that "in most families with two cohabiting spouses . . . the eighteenth century saw a narrowing of women's economic roles, a decline in their public economic activities." Kathleen Brown maintained that in eighteenth-century Virginia in particular, men divided women into "good wives" and "nasty wenches," both of whom lived in ever-growing fear of male violence. In these histories, eighteenth-century women lost value rapidly. See Cornelia Hughes Dayton, *Women before the Bar: Gender, Law, and Society in Connecticut, 1639–1789* (Chapel Hill: published for the IEAHC by the University of North Carolina Press, 1995); Deborah A. Rosen, *Courts and Commerce: Gender, Law, and the Market Economy in Colonial New York* (Columbus: Ohio State University Press, 1997), 111; Jeanne Boydston, *Home and Work: Housework, Wages, and the Ideology of Labor in the Early Republic* (New York: Oxford University Press, 1990), xi, 18; Cynthia A. Kierner, *Beyond the Household: Women's Place in the Early South, 1700–1835* (Ithaca, NY: Cornell University Press, 1998), 13.

CHAPTER ONE: "IT WAS BEING TOO ABSTEMIOUS
THAT BROUGHT THIS SICKNESS UPON ME"

1. Small ale was the liquor made when water was pushed through the mash a second time.

2. Philip Vickers Fithian, *Journal 1775–1776: Written on the Virginia-Pennsylvania Frontier and in the Army around New York*, ed. Robert Albion and Leonidas Dodson (Princeton: Princeton University Press, 1934), 121; Rorabaugh, *Alcoholic Republic*, 8, 10.

3. Peter Clark, *The English Alehouse: A Social History 1200–1830* (New York: Longman, 1983), 32, 112; Christina Hole, *The English Housewife in the Seventeenth Century* (London: Chatto & Windus, 1953), 60, 84–85.

4. Clark, *English Alehouse*, 112; Andrew Barr, *Drink: A Social History of America* (New York: Carroll & Graf, 1999), 31–32.

5. A. Lynn Martin, "How Much Did They Drink? The Consumption of Alcohol in Traditional Europe," email to Alcohol History and Temperance Society listserv, 3 January 2001; "One Day in America," *Time*, 22 October 2008.

6. Elizabeth Freke, "Diary," in *"Capacious Hold-All": An Anthology of English-women's Diary Writings*, ed. Harriet Blodgett (Charlottesville: University of Virginia Press, 1991), 37. David Underdown, *Fire From Heaven: Life in an English Town in the Seventeenth Century* (London: HarperCollins, 1992), 114.

7. Martin, "How Much Did They Drink?" n.p.

8. Eliza Haywood, *A Present for a Servant-Maid* (London, 1743), 6–7, LCP; Anon., *The Servants' Guide and Family Manual*, 2nd ed. (London, 1831), microfilm, Alderman Library, 118; Lydia Maria Child, *The Family Nurse; or, Companion of the Frugal Housewife* (Boston, 1837), microfilm, Alderman Library, 9; Barr, *Drink*.

9. Kenelme Digbie, *The Closet of the Eminently Learned Sir Kenelme Digbie Kt.*, ed. Jane Stevenson and Peter Davidson (Boston: Prospect Books, 1997), xxx.

10. Clark, *English Alehouse*, 112; Liza Picard, *Restoration London: From Poverty to Pets, from Medicine to Magic, from Slang to Sex, from Wallpaper to Women's Rights* (New York: St. Martin's, 1998), 93; Freke, "Diary," 32, 33.

11. Freke, "Diary," 33, 107; *The Autobiography of Mrs. Alice Thornton of East Newton, Co. York*, ed. Charles Jackson (published for the Surtees Society by Andrews and Co., 1875), 388; Brian Harrison, *Drink and the Victorians: The Temperance Question in England 1815–1871* (Pittsburgh: University of Pittsburgh Press, 1971), 39.

12. Edward Pond, *An Almanack for the Year Our Lord God 1708* (London, 1707), n.p., LCP.

13. Haywood, *Present for a Servant-Maid*, 6–7; Barr, *Drink*, 39; Anon., *Servants' Guide and Family Manual*, 118.

14. Gervase Markham, *Country Contentments, or The English Huswife* (London, 1623), microfilm, Alderman Library, 11; *Servants' Guide and Family Manual*, 118; Richard Leveridge, "This Great World is a Trouble," 1723 in John Edmunds, *A Williamsburg Songbook: Songs, Convivial, Sporting, Amorous, &c. from Eighteenth-Century*

Collections known to have been in the Libraries of Colonial Virginians (New York: Holt, Rinehart and Winston for Colonial Williamsburg, 1964), 4.

15. Hole, *English Housewife*, 71–72.

16. Robert Roberts, *The House-Servant's Directory; or, a Monitor for Private Families*, 2nd ed. (Boston, 1828), microfilm, Alderman Library, 20–21, 24, 80, 87, 92, 33, 108, 27.

17. Barr, *Drink*, 44; Hole, *English Housewife*, 59.

18. Henry Hartwell, James Blair, and Edward Chilton, *The Present State of Virginia, and the College*, ed. Hunter Dickinson Farish (Charlottesville: University Press of Virginia, 1940), 9–10; Edwin J. Perkins, *The Economy of Colonial America* (New York: Columbia University Press, 1980), 48; Gregory A. Stiverson and Patrick H. Butler III, eds., "Virginia in 1732: The Travel Journal of William Hugh Grove," *VMHB* 85, no. 1 (1977): 24; Rorabaugh, *Alcoholic Republic*, 98–99.

19. Percy qtd. in Steven G. Davison et al., *Chesapeake Waters: Four Centuries of Controversy, Concern and Legislation*, 2nd ed. (Centreville, MD: Tidewater Publishers, 1997), 3; Barr, *Drink*, 32; Wyndham B. Blanton, *Medicine in Virginia in the Seventeenth Century* (Richmond, VA: William Byrd Press, 1930), 54–55.

20. Barr, *Drink*, 33.

21. The first effort to protect drinking water in the Chesapeake was not until 1792 when the Insurance Fire Company of Baltimore unsuccessfully attempted to supply the town with a reservoir and pipes to conduct water. In 1808 the Baltimore Water Company made a similar attempt, but only for part of the town, and the resulting water was a "muddy substitute for the pure element," citizens complained. Gloria Main, *Tobacco Colony: Life in Early Maryland 1650–1720* (Princeton: Princeton University Press, 1982), 197; Blanton, *Medicine in Virginia*, 72; Davison et al., *Chesapeake Waters*, 35; Moses N. Baker, *The Quest for Pure Water: The History of Water Purification from the Earliest Records to the Twentieth Century* (New York: American Water Works Association, 1948), 35–36; Rorabaugh, *Alcoholic Republic*, 97.

22. *Caroline County, Virginia Order Book 1759–1763: Part One 1759–1760*, ed. John Frederick Dorman, 15 June 1759 (n.p., 1982); Nicholas Cresswell, *The Journal of Nicholas Cresswell 1774–1777*, ed. Lincoln MacVeagh (New York: Dial Press, 1924), 22; St. Mery, qtd. in Barr, *Drink*, 37; Theophilus Wreg, *The Virginia Almanck for the Year of our Lord God 1766* (Williamsburg, 1765), n.p., LCP.

23. Barr, *Drink*, 34.

24. Barr, *Drink*, 34, 36; Lucious Verus Bierce, *Travels in the Southland 1822–1823: The Journal of Lucious Verus Bierce*, ed. George W. Knepper (Columbus: Ohio State University Press, 1966), 53, 64.

25. *The Diary of Colonel Landon Carter of Sabine Hall, 1752–1778*, ed. Jack P. Greene (Richmond: Virginia Historical Society, 1965), 1:177, 2:635, 1:321.

26. Hannah Huthwaite Recipe Book, ca. 1720, MS, Winterthur Library; Bierce, *Travels in the Southland*, 56; *Diary of Landon Carter*, 1:321; "Journal of Col. James Gordon of Lancaster County, Va.," 5 April 1760, *WMQ* 11, no. 3 (Jan. 1903): 198; Barr, *Drink*, 34–35.

27. Child, *Family Nurse*, 5; William Mylne to Robert Mylne, 26 June 1774, in *Trav-*

els in the Colonies 1773–1775: Described in the Letters of William Mylne, ed. Ted Ruddock (Athens: University of Georgia Press, 1993), 33.

28. John Blackford, *Ferry Hill Plantation Journal January 4, 1838 to January 15, 1839*, ed. Fletcher M. Green (Chapel Hill: University of North Carolina Press, 1961), 65; *Diary of Landon Carter*, 19 April 1770.

29. Salinger, *Taverns and Drinking in Early America*, 67–68; Francis Louis Michel, "Report of the Journey of Francis Louis Michel from Berne, Switzerland, to Virginia October 2, 1701 – December 1, 1702," trans. and ed. William J. Hinke, *VMHB* 24, no. 2 (1916): 127; William Byrd, *The Secret Diary of William Byrd of Westover 1709–1712*, ed. Louis B. Wright and Marion Tinling (Richmond, VA: Dietz Press, 1941), 270, 298; John Fontaine, *The Journal of John Fontaine, An Irish Huguenot Son in Spain and Virginia 1710–1719*, ed. Edward Porter Alexander (Williamsburg: Colonial Williamsburg Foundation, 1972), 106; *Maryland Gazette*, 17 June 1746.

30. Byrd, *Secret Diary*, 324; Cynthia A. Kierner, *Beyond the Household: Women's Place in the Early South, 1700–1835* (Ithaca, NY: Cornell University Press, 1998), 34; William Byrd, *Histories of the Dividing Line betwixt Virginia and North Carolina*, ed. William K. Boyd (Raleigh: North Carolina Historic Commission, 1929), 21.

31. *Lower Norfolk County, Virginia, Court Records: Book "A" 1637 to 1746 and Book "B" 1646–1652*, 16 Dec. 1647, ed. Alice Granbery Walter (Baltimore: Clearfield Company, 1994), 58; Byrd, *Secret Diary*, 218; *Accomack County, Virginia Court Order Abstracts 1673–1676*, 19 Nov. 1675, ed. Joann Riley McKey (Bowie, MD: Heritage Books, 1997), 4:151, 9:vi, 68; Byrd, *Secret Diary*, 98; A. G. Roeber, *Faithful Magistrates and Republican Lawyers: Creators of Virginia Legal Culture, 1680–1810* (Chapel Hill: University of North Carolina Press, 1981), 81–82.

32. Byrd, *Secret Diary*, 232, 234; Robert Wormeley Carter, *Diary*, 30 July 1769, Rockefeller Library.

33. Rorabaugh, *Alcoholic Republic*, 14, 20; David Rittenhouse, "Gunpowder: Of What It Consists and How It Is Made," *The Virginia Almanack for the Year of our Lord God 1776* (Williamsburg, 1775), n.p., LCP; Janet Schaw, *Journal of a Lady of Quality: Being the Narrative of a Journey from Scotland to the West Indies, North Carolina, and Portugal, in the Years 1774 to 1776*, ed. Evangeline Walker Andrews (New Haven: Yale University Press, 1922), 190; E. Wayne Carp, *To Starve the Army at Pleasure: Continental Army Administration and American Political Culture 1775–1783* (Chapel Hill: University of North Carolina Press, 1984), 55.

34. Thomas R. Pegram, *Battling Demon Rum: The Struggle for a Dry America, 1800–1933* (Chicago: Ivan R. Dee, 1998), ix; Robert Munford, "The Candidates," ed. Jay B. Hubbell and Douglass Adair, *WMQ*, 3rd ser., 5, no. 2 (April 1948): 229; David John Mays, *Edmund Pendleton 1721–1803* (1952; Richmond: Virginia State Library, 1984), 1:122; Moreau de St. Mery, *American Journey [1793–1798]*, trans. and ed. Kenneth Roberts and Anna M. Roberts (Garden City, NY: Doubleday, 1947), 90.

35. Anon., "Trade and Travel in Post-Revolutionary Virginia: A Diary of an Itinerant Peddler, 1807–1808," ed. Richard R. Beeman, *VMHB* 84, no. 2 (1976): 183; *Virginia Gazette*, Purdie and Dixon, 20 Aug. 1772.

36. Frances Baylor Hill, "The Diary of Frances Baylor Hill of Hillsborough, King

and Queen County, Virginia (1797)," ed. William K. Bottorff and Roy Flannagan, *Early American Literature Newsletter* 2 (1967): 38; Byrd, *Secret Diary*, 60; Deborah Gray White, *"Ar'n't I a Woman?": Female Slaves in the Plantation South* (New York: W.W. Norton, 1985), 111; refrain of the song "Lucretia" possibly by Edward Betts, ca. 1724, in Edmunds, *A Williamsburg Songbook*, 31.

37. Anon., "Observations in Several Voyages and Travels in America," WMQ 16, no. 1 (1907): 4; Byrd, *Secret Diary*, 394; John Harrower, *The Journal of John Harrower: An Indentured Servant in the Colony of Virginia, 1773–1776*, ed. Edward Miles Riley (New York: Holt, Rinehart and Winston for Colonial Williamsburg, 1963), 7, 73, 126; Harry Toulmin, *The Western Country in 1793: Reports on Kentucky and Virginia*, ed. Marion Tinling and Godfrey Davies (San Marino, CA: Printed by Castle Press for the Henry E. Huntington Library and Art Gallery, 1948), 25; Isaac Weld, *Travels through the States of North America* (1807; New York: Johnson Reprint Corporation, 1968), 187.

38. Byrd, *Secret Diary*, 165; qtd. in Salinger, *Taverns and Drinking*, 67–68, 70.

39. Funeral and holiday quotes from Philip Morgan, *Slave Counterpoint: Black Culture in the Eighteenth-Century Chesapeake and Lowcounty* (Chapel Hill: Published for the OIEAHC by the University of North Carolina Press, 1998), 413; Harrower, *Journal of John Harrower*, 79; Susan Dabney Smedes, *Memorials of a Southern Planter* (Baltimore, 1886), 60, 58.

40. Thomas Jefferson, *Memorandum Books: Accounts, with Legal Records and Miscellany, 1767–1826*, 25 June 1813, ed. James A. Bear and Lucia C. Stanton (Princeton: Princeton University Press, 1997); Lawrence William McKee, "Plantation Food Supply in Nineteenth-Century Tidewater Virginia" (Ph.D. diss., University of California, Berkeley, 1988), 128; Green, *Ferry Hill*, 2 Aug. 1838; Smedes, *Memorials*, 56; Edwin J. Perkins, *The Economy of Colonial America* (New York: Columbia University Press, 1980), 74.

41. Smedes, *Memorials*, 83.

42. Joseph Ball to Joseph Chinn, 25 March 1743, "Joseph Ball Letterbook 1743–1780," qtd. in Edward Ayres, "Fruit Culture in Colonial Virginia," *Colonial Williamsburg Early American History Research Report* (1973), microfiche, Alderman Library, 142; Smedes, *Memorials*, 66; Ralph Wormeley Carter, *Diary*, 20 Feb. 1785; qtd. in Morgan, *Slave Counterpoint*, 641.

43. Morgan, *Slave Counterpoint*, 641; *Diary of Landon Carter*, 1:484; Julian Ursyn Niemcewicz, *Under Their Vine and Fig Tree: Travels through America in 1797–1799, and 1805*, ed. Metchie J. E. Budka (Elizabeth, NJ: New Jersey Historical Society at Newark by Grassman Publishing, 1965), 101.

44. Byrd, *Secret Diary*, 56, 42, 53; Col. James Tayloe to Landon Carter, 31 March 1771, Landon Carter Papers, MSS, Alderman Library.

45. McKee, "Plantation Food Supply," 138; Jane Carson, "Plantation Management," *Colonial Williamsburg Research Report* (1974), microfiche, Alderman Library, 41–42; Morgan, *Slave Counterpoint*, 365; Fauntleroy to Landon Carter, 10 July 1774, Landon Carter Papers, MSS, Alderman Library.

46. Peter Charles Hoffer and William B. Scott, eds., *Criminal Proceedings in Colonial Virginia: [Records of] Fines, Examination of Criminals, Trials of Slaves, etc., from*

March 1710 [1711] to [1754] [Richmond County, Virginia] (Athens: published for the American Historical Association by the University of Georgia Press, 1984), 42; Morgan, *Slave Counterpoint*, 415; Byrd, *Secret Diary*, 337.

47. Morgan, *Slave Counterpoint*, 415; White, *"Ar'n't I a Woman?"* 138; Ann Smart Martin, "Buying into the World of Goods: Eighteenth-Century Consumerism and the Retail Trade from London to the Virginia Frontier" (Ph.D. diss., College of William & Mary, 1993), 309; qtd. in Morgan, *Slave Counterpoint*, 433–34.

48. Lorena S. Walsh, Ann Smart Martin, Joanne Bowen, with contributions by Jennifer A. Jones and Gregory J. Brown, *Provisioning Early American Towns. The Chesapeake: A Multidisciplinary Case Study* (http://research.history.org/Archaeological_Research/Research_Articles/ThemeZooarch/Provisioning.cfm, 1997), 140; Rorabaugh, *Alcoholic Republic*, 7–10.

49. Robert D. Mitchell, *Commercialism and Frontier: Perspectives on the Early Shenandoah Valley* (Charlottesville: University Press of Virginia, 1977), 141; qtd. in McKee, "Plantation Food Supply," 48, 43.

CHAPTER TWO: "THEY WILL BE ADJUDGED BY THEIR DRINKE,
WHAT KIND OF HOUSEWIVES THEY ARE"

1. John Hammond, "Leah and Rachel, or, The Two Fruitfull Sisters, Virginia and Mary-land," in *Narratives of Early Maryland 1633–1684*, ed. Clayton Colman Hall (New York: Charles Scribner's Sons, 1910), 292.

2. Peter Charles Hoffer and William B. Scott, eds., *Criminal Proceedings in Colonial Virginia: [Records of] Fines, Examination of Criminals, Trials of Slaves, etc., from March 1710 [1711] to [1754] [Richmond County, Virginia]* (Athens: University of Georgia Press, 1984), 43.

3. "Letters of the Byrd Family," VMHB 37, no. 2 (1929): 109; John Custis to Mr. Loyd [or Boyd] 1733, *John Custis Letterbook 1717–1744*, 2:131, TS, Rockefeller Library; Byrd, *Secret Diary*, 315, 211.

4. Continental winemaking was generally men's domain, although widows sometimes continued their husband's wine-making trade. English women made household "wines," which did not necessarily ferment, from a variety of garden berries and fruits. Susan Cahn, *Industry of Devotion: The Transformation of Women's Work in England 1500–1660* (New York: Columbia University Press, 1987), 33; Clark, *English Alehouse*, 47; Thomas Tusser, *Thomas Tusser 1557 Floruit: His Good Points of Husbandry*, ed. Dorothy Hartley (New York: Augustus M. Kelley Publishers, 1970), 166, 169, 171; Hole, *English Housewife*, 108–109; Stanley Baron, *Brewed in America: A History of Beer and Ale in the United States* (Boston: Little, Brown, 1962), 14.

5. Richard Bradley, *The Country Housewife and Lady's Director* (1727; London, 1732), vii, 51–52, LCP.

6. Mary Cole, *The Lady's Complete Guide; or, Cookery in all its Branches*, 2nd ed. (London, 1791), 361, LCP; Digbie, *The Closet*, 81.

7. By 1400 Holland and Germany already had highly masculinized beer brewing

industries with breweries of the size that England would not reach for another one hundred years. Hole, *English Housewife*, 58–59.

8. John Burnett, *Liquid Pleasures: A Social History of Drinks in Modern Britain* (New York: Routledge, 1999), 157; Bennett, *Ale, Beer, and Brewsters*, 11–12; Thomas Chapman, *The Cyder-Maker's Instructor, Sweet-Makers Assistant, and Victualler's and Housekeeper's Director* (London, 1762), 9, LCP.

9. Francis Wyatt, 3 April 1623, qtd. in Baron, *Brewed in America*, 6; David R. Ransome, "Wives for Virginia, 1621," *WMQ*, 3rd ser., 48, no. 1 (1991): 15.

10. Lois Green Carr and Lorena S. Walsh, "The Planter's Wife: The Experience of White Women in Seventeenth-Century Maryland," in Nancy F. Cott and Elizabeth H. Pleck, *A Heritage of Her Own: Toward a New Social History of American Women* (New York: Simon and Schuster, 1979), 26–29.

11. Stratton Nottingham, *Land Causes of Accomack County Virginia, 1727–1826* (1930; Baltimore: Genealogical Publishing, 1999), 5; Weynette Parks Haun, *Surry County, Virginia, Court Records, 1664 thru 1671* (Durham, NC: n.p., 1987), 2:44; Eliza Timberlake Davis, *Wills and Administrations of Surry County, Virginia 1671–1750* (Baltimore: Clearfield, 1995), 9; Darrett and Anita Rutman, *A Place in Time: Middlesex County, Virginia, 1650 to 1750* (New York: Norton, 1984), 106; "Will of Richard Wharton, 1713, Williamsburg," *Virginia Colonial Records Project*, LVA, http://lvaimage .lib.va.us/VTLS/CR/04359/index.html Survey Report Image; Charles City County, Virginia, Order Book, Aug. 1722 to Feb. 1722/3, MSS, VHS, 143, 151; John Banister, "John Banister Account Book 1731–1734," MS, VHS, n.p.; "Part of Grymes' Estate Divided," Middlesex County Court, 4 July 1749, *VMHB* 27, nos. 3 and 4 (July and Oct. 1919): 406.

12. Will of Elizabeth Ballard, 1 Oct. 1726, Inventory of Sabra Crew, 19 April 1726, Inventory of Mary Tye, 1 Nov. 1729, in Will Book, 1724/5 February 3 — 1731 June 2, *Charles City County*, photocopy, VHS.

13. Inventory of Ambrose Fielding, 17 March 1674, *VMHB* 15, no. 1 (July 1902): 206; "Room-by-room inventories 1646–1824," MS, Rockefeller Library. Inventory of the Estate of Nathaniel Bradford, Accomack County, deceased 1690 (no further date); Inventory of John Thompson, Surry County, 18 June 1700; Inventory of James Burwell, York County, 10 May 1718; Inventory of Joseph Ring, York County, 1704 (no further date); Inventory of Joseph Frith, York County, 16 June 1712; Inventory of Madam Katherine Gwyn, Richmond County, 6 Nov. 1728; Inventory of Arthur Allen, Surry County, May and June 1711 (no further date).

14. Letitia M. Burwell, *A Girl's Life in Virginia before the War* (New York, 1895), 33; Mary Randolph, *The Virginia Housewife; or, Methodical Cook* (Philadelphia, 1846); Catherine Clinton, *The Plantation Mistress: Woman's World in the Old South* (New York: Pantheon Books, 1982), 28; Hannah Huthwaite, *Hannah Huthwaite Recipe Book* (ca. 1720), MS, Winterthur Library; Frances Parke Custis Cook Book, photocopy, CWF, published as Martha Washington, *Martha Washington's Booke of Cookery and Booke of Sweetmeats: Being a Family Manuscript*, transcribed by Karen Hess (New York: Columbia University Press, 1981).

15. Lois Green Carr, Russell R. Menard, and Lorena S. Walsh, *Robert Cole's*

World: Agriculture and Society in Early Maryland (Chapel Hill: published for the OIEAHC by the University of North Carolina Press, 1991), 72–74.

16. Hugh Plat, *Divers Chimicall Conclusions Concerning the Art of Distillation* (London, 1594), Dibner Library; Peter Kalm, *Travels in North America 1748–1749*, trans. John Reinhold Forster (reprint, Barre, MA: Imprint Society 1972), 2:67, 73–74.

17. St. John De Crevecoeur, *Sketches of Eighteenth Century America*, ed. Henri L. Bourdin, Ralph H. Gabriel, and Stanley T. Williams (New Haven: Yale University Press, 1925), 124, 40; James Fordyce, "Sermons for Young Women," in *Female Education in the Age of the Enlightenment*, ed. Janet Todd (London: William Pickering, 1996), 1:214; Benjamin Franklin, *Poor Richard's Almanack for the Year of Our Lord 1746* (1745; New York: David McKay, 1976), 81.

18. James Horn, *Adapting to a New World: English Society in the Seventeenth-Century Chesapeake* (Chapel Hill: published for the IEAHC by the University of North Carolina Press, 1994), 25, 39–41.

19. Carr, Menard, and Walsh, *Robert Cole's World*, 55–76.

20. Baron, *Brewed in America*, 32. See Royal Dublin Society, *Instructions for Planting and Managing Hops and for Raising Hop-Poles* (Dublin, 1733), 8–9, 12, Winterthur Library.

21. For laws encouraging men to grow hops, see, for example, W. W. Henning, *The Statutes at Large: Being a Collection of all the Laws of Virginia from the First Session of the Legislature in the Year 1619*, Act LXXVI (Richmond, 1809), 1:469.

22. Harry Toulmin, *The Western Country in 1793: Reports on Kentucky and Virginia*, ed. Marion Tinling and Godfrey Davies (San Marino, CA: Printed by Castle Press for the Henry E. Huntington Library and Art Gallery, 1948), 45, 63.

23. H. S. Corran, *A History of Brewing* (Newton Abbot, UK: David and Charles, 1975), 273–74. Martha J. Randolph to Ellen Wayles Randolph Coolidge, *Jefferson Papers*, 26 Nov. 1825, mss., Alderman Library; Jean B. Russo "Self-Sufficiency and Local Exchange: Free Craftsmen in the Rural Chesapeake Economy," in Lois Green Carr, Philip D. Morgan, and Jean B. Russo, eds., *Colonial Chesapeake Society* (Chapel Hill: published for the OIEAHC by the University of North Carolina Press, 1988), 388–432, 394; *Maryland Gazette*, 18 Feb. 1746.

24. Bennett, *Ale, Beer, and Brewsters*, 17.

25. Robert Beverley, *The History and Present State of Virginia*, ed. Louis B. Wright (Charlottesville, NC: Dominion Books, 1947), 293; "Observations," 3.

26. Horn, *Adapting to a New World*, 434.

27. Durand de Dauphine, *A Huguenot Exile in Virginia: or Voyages of a Frenchman Exiled for his Religion with a Description of Virginia and Maryland*, ed. Gilbert Chinard (1687; New York: Press of the Pioneers, 1934); "Observations," 2; Peter Collinson to John Custis, believed to be 3 July 1735, in *Brothers of the Spade: Correspondence of Peter Collinson of London, and of John Custis, of Williamsburg, Virginia, 1736–1746*, ed. E. G. Swem (Barre, MA: Barre Gazette, 1957), 28; Hazard, "Journal of Ebenezer Hazard," 420; Eugene Genovese, *Roll Jordan Roll: The World the Slaves Made* (New York: Vintage Books, 1972), 644; Thomas Campbell, Interview of Sarah Fitzpatrick, in *Slave Testimony: Two Centuries of Letters, Speeches, Interviews, and*

Autobiographies, ed. John W. Blassingame (Baton Rouge: Louisiana State University Press, 1977), 639–55.

28. Anon., "Trade and Travel in Post-Revolutionary Virginia," 178; Beverley, *History and Present State of Virginia*, 293.

CHAPTER THREE: "THIS DRINK CANNOT BE KEPT DURING THE SUMMER"

1. Small, middling, and large planters are here defined as Aubrey C. Land classified them in "Economic Base and Social Structure: The Northern Chesapeake in the Eighteenth Century," in *Shaping Southern Society: The Colonial Experience*, ed. T. H. Breen (New York: Oxford University Press, 1977), 237–38.

2. Harry Toulmin, *The Western Country in 1793: Reports on Kentucky and Virginia*, ed. Marion Tinling and Godfrey Davies (San Marino, CA: Castle Press for the Henry E. Huntington Library and Art Gallery, 1948), 25; Anon., "Observations in Several Voyages and Travels in America," *WMQ* 16, no. 1 (1907): 3–4.

3. *Virginia Gazette*, 3 Sept. 1736; Beverley, *History and Present State of Virginia*, 314; "An Act to prevent horses running at large and barkeing fruit trees," in William Walter Hening, *The Statutes at Large: Being a Collection of all the Laws of Virginia, from the First Session of the Legislature in the year 1619* (New York: 1819–1823), 3:70–71 (laws published in 1691, 1733, and 1752).

4. Greene, *Diary of Colonel Landon Carter*, 2:589. See, for example, Crèvecoeur, *Sketches of Eighteenth Century America*, 86, 121, 124, 40; Edward Ayres, "Fruit Culture in Colonial Virginia," in *Colonial Williamsburg Early American History Research Report* (1973), microfiche, Alderman Library, 122; Johann David Schhoepf, *Travels in the Confederation 1783–1784*, trans. Alfred J. Morrison (Philadelphia: William J. Campbell, 1911), 2:38–39; Philip Vickers Fithian, *Journal and Letters of Philip Vickers Fithian 1773–1774: A Plantation Tutor of the Old Dominion*, ed. Hunter Dickinson Farish, 4 June 1774 (Williamsburg, VA: Colonial Williamsburg,1957).

5. Robert Carter to James Buchanan and Co., 10 May 1764, in "Letterbook of Robert Carter III of Nomini Hall for the Years 1764–1768," MS, Rockefeller Library; John Custis to Peter Collinson, believed to be 12 Aug. 1739, in *Brothers of the Spade*, 61, 166, n. 111; Darret and Anita Rutman, *A Place in Time: Middlesex County, Virginia 1650–1750* (New York: W.W. Norton, 1984), 213.

6. "Ledger of Imports and Exports, Christmas 1724 to Christmas 1725" (PRO Customs 3/25), *Virginia Colonial Records Project* microfilm roll 210, Alderman Library; "Ledger of Imports and Exports, Christmas 1734 to Christmas 1735" (PRO Customs 3/35), VCRP roll 210; "Ledger of Imports and Exports, Christmas 1744 to Christmas 1745" (PRO Customs 3/45), VCRP roll 210; "Ledger of Imports and Exports, Christmas 1754 to Christmas 1755" (PRO Customs 3/55), VCRP roll 232; "Ledger of Imports and Exports, Christmas 1764 to Christmas 1765" (PRO Customs 3/65), VCRP roll 233; "Ledger of Imports and Exports, Christmas 1774 to Christmas 1775" (PRO Customs 3/75), VCRP roll 233.

7. Robert Bennett to Edward Bennett, 9 June 1623, in *The Records of the Virginia Company of London*, ed. Susan Myra Kingsbury (Washington, DC: Government

Printing Office, 1906–35), 4:220; Robert Carter to Edward Tucker, 13 July 1720, in *Letters of Robert Carter 1720–1727: The Commercial Interests of a Virginia Gentleman*, ed. Louis B. Wright (Westport, CT: Greenwood Press, 1970), 16.

8. Richard Corbin, "Richard Corbin Letterbook, 1758–1768," 15 June 1758, microfilm, LVA, 15; Colonel James Gordon, "Journal of Col. James Gordon of Lancaster County (VA) 1758–1768," 2 Feb. and 7 April 1760, ms., VHS; John Custis to Robert Cary, 1729, "John Custis Letterbook," TS, 93, Rockefeller Library; Carter, *Letters of Robert Carter*, 30–31; Custis to Mr. Perry, 1722, 36; to Loyd and Cooper, 1736; to Loyd and Cooper 1737; to Johnathon Day 1736; to Robert Cary 1738, "John Custis Letterbook," 163, 192, 175, 199; "Letters of the Byrd Family," *VMHB* 37, no. 2 (1929): 110.

9. Custis to Robert Cary, 1739, "John Custis Letterbook," 210; Robert Carter III to Francis Fauquier, "Letterbook of Robert Carter III," 23 May 1769, MS, Rockefeller Library, 10; Byrd, *Secret Diary*, 398; *Diary of Colonel Landon Carter*, 2:67; Custis to Johnathon Day, 10 July 1741, in "John Custis Letterbook," 175; "Letters of the Byrd Family," *VMHB* 37, no. 2 (1929): 110.

10. Carter Burwell, "Littletown Plantation Ledger, 1736–1746," in *Burwell Family Papers, 1736–1810*, microfilm, Rockefeller Library.

11. James P. McClure, "Littletown Plantation, 1700–1745" (master's thesis, College of William and Mary, 1977), 10–15, 17–19, 29, 49, 51, 85, 98.

12. Ibid., 34–37.

13. Ibid., 58.

14. John Carter to Micajah Perry, 31 Aug. 1736 and John Carter to John Hanbury, 31 Aug. 1738, quoted in Mary A. Stephenson, *Carter's Grove Plantation: A History* (Williamsburg, VA: Prepared for Sealantic Fund by Research Department, Colonial Williamsburg, 1964), 248, 249.

15. Robert Polk Thomson, "The Merchant in Virginia 1700–1775" (Ph.D. diss., University of Wisconsin, 1955), 161; Michel, "Report of the Journey," 140.

16. Robert Wormeley Carter, "Diary 1765–1792," 3 Dec. 1765, 11 Oct. 1774, 2 Jan. 1777, 19 July 1792, 4 July 1792, 9 July 1769, 6 Aug. 1790, TS, Rockefeller Library.

17. McClure, "Littletown Plantation," 84, 88.

18. Walsh et al., *Provisioning Early American Towns*, 16–17; Stephenson, *Carter's Grove Plantation*, 46–47; Byrd, *Secret Diary*, 488; *Major-Marable family papers*, 1703–1929, microfilm, Rockefeller Library.

19. McClure, "Littletown Plantation," 77–78, 80–81, 83, 85, 91–92.

20. Ibid., 96–98.

21. Gerald F. Moran and Maris A. Vinovskis, "Literacy and Education in Eighteenth-Century North America," in *The World Turned Upside-Down: The State of Eighteenth-Century American Studies at the Beginning of the Twenty-First Century*, ed. Michael V. Kennedy and William G. Shade (Bethlehem, PA: Lehigh University Press, 2001), 191–92; Robert Wormeley Carter, "Diary 1769," 1 Jan. 1769, TS, Rockefeller Library, 2; See Greene, *Diary*, 529, for an example of Landon Carter recording dairy purchases; Heather R. Wainwright, "Inns and Outs: Anne Pattison's Tavern Account Book, 1744–1749" (master's thesis, Armstrong Atlantic State University, 1998), 86; *Virginia Gazette*, 29 Aug. 1766.

22. Fithian, *Journal and Letters*, 114, 175; Robert Wormeley Carter, "Diary," 29 Dec. 1765, TS, Rockefeller; Landon Carter, *Diary*, 2:1144–45; Jane Carson, "Plantation Management," Colonial Williamsburg Research Report (1974), 41–42, microfiche, Alderman Library; Thomas Jefferson, *The Garden and Farm Book*, ed. Robert C. Baron (Golden, CO: Fulcrum, 1987), 268; Walsh et al., *Provisioning Early American Towns*, 88; Byrd, *Secret Diary*, 472; 388; "Journal of Col. James Gordon of Lancaster County, Virginia, 1758–1768," 3 Sept. 1758, MS, Virginia Historical Society; George Washington, *Weekly Farm Reports 1786–1797*, MS, Mount Vernon Library.

23. Frederick H. Smith, *Caribbean Rum: A Social and Economic History* (Gainesville: University of Florida, 2005).

24. William Augustine Washington was the son of Washington's half brother. These examples are admittedly chronologically late for this chapter, but they are the earliest known references. William Augustine Washington to George Washington, 1 June 1785, *The Papers of George Washington Digital Edition*, ed. Theodore J. Crackel (Charlottesville: University of Virginia Press, Rotunda 2007), http://rotunda.upress .virginia.edu/pgwde/search-Con03d25; Niemcewicz, *Under Their Vine and Fig Tree*, 100; Lund Washington to George Washington, 17 Aug. 1767; Lund Washington to George Washington, 5 Sept. 1767; George Washington and Nelson Kelly, Agreement with Nelson Kelly, 1 Sept. 1762, *Papers of George Washington Digital Edition*.

25. Ayres, "Fruit Culture," 214; Thomas Glover, "An Account of Virginia" (1676), 13–14, microfilm, LC; Richard Allen to Jonathon Owen, "Miscellaneous Colonial Documents," 14 July 1704, *VMBH* 19, no. 1 (1911): 12–13.

26. An apple tree produced from one to twenty bushels. A typical eight-year-old apple tree gave two bushels of apples a year, providing Fitzhugh with an estimated 5,000 bushels of apples. One hogshead of cider (between forty-eight and sixty-three gallons) required twenty to thirty bushels of apples a year. Using the median number of bushels, Fitzhugh's plantation could make at least two hundred hogsheads, or a minimum of 9,600 gallons of cider per year. William Fitzhugh to Ralph Smith, 22 April 1686, in *William Fitzhugh and His Chesapeake World 1676–1701: The Fitzhugh Letters and Other Documents*, ed. Richard Beale Davis (Chapel Hill: published for the Virginia Historical Society by the University of North Carolina Press, 1963), 175.

27. See Ayres, "Fruit Culture," 141, 142; Stephenson, *Carter's Grove Plantation*, 7; Robert Wormeley Carter, "Diary 1768," TS, Rockefeller Library; Robert Wormeley Carter, "Diary 1776," 13 July 1776, TS, Rockefeller Library; Thomas Mallory to John Norton and Sons, 5 Sept. 1771, in *John Norton & Sons, Merchants of London and Virginia: Being the Papers from their Counting House for the Years 1750 to 1795*, ed. Frances Norton Mason (Richmond, VA: Dietz Press, 1937); Greene, *Diary*, 2:931, 932, 192, 1118, 1135, 1149; "Inventory of Estate of Landon Carter esq.," Feb. 1779, MS, VHS; Greene, *Diary*, 2:534; Landon Carter, "1779 Essay on The Cultivation of Hops," MS, Landon Carter Papers, Alderman Library; Daniel Roberdeau, "Letterbook 1767–1791," MS, LC, 123, 184, 95, 216, 183.

28. Isaac Weld, *Travels through the States of North America* (1807; New York: Johnson Reprint Corporation, 1968), 182–83; Walsh et al., *Provisioning Early American Towns*, 12.

29. Ayres, "Fruit Culture," 141, 145–47; 143; Stephenson, *Carter's Grove Plantation,* 251; *Virginia Gazette,* 17 June 1737.

30. Robert Wormeley Carter, "Diary," 13 July 1780, TS, Rockefeller Library, 17; Washington qtd. in Gregory A. Stiverson, "Gentlemen of Industry, Skill, and Application: Plantation Management in Eighteenth-Century Virginia," *Colonial Williamsburg Early American History Research Report* (1975), microfiche, Alderman Library, 21–22; Richard Henry Lee to Mr. Valentine, 20 Jan. 1783, in "Some Notes on Green Spring," *VMHB* 38, no. 1 (1930): 39; William Fitzhugh to Nicholas Hayward, 20 May 1691, in *William Fitzhugh,* 291; Beverley, *History and Present State of Virginia,* 314.

31. Ayres, "Fruit Culture," 33, 127, 14, 12–13; William Byrd to Mr. Warner, July 1729, "Letters of the Byrd Family," 116; Lee to Mr. Valentine, "Some Notes on Green Spring," 20 Jan. 1783, *VMHB* 38, no. 1 (Jan. 1930): 39.

32. Robert Wormeley Carter, "Diary," 20 Feb. 1768, 27 Feb. 1768.

33. John Worlidge, *Vinetum Britannicum: or a Treatise of Cider, and other Wines and Drinks extracted from Fruits Growing in this Kingdom* (London, 1678), 27, 31, Winterthur; "Joseph Ball Letterbook, 1743–1780," MS, LC, 119, 121; *The Commonplace Book of William Byrd II of Westover,* ed. Kevin Berland, Jan Kirsten Gilliam, and Kenneth A. Lockridge (Chapel Hill: published for the OIEAHC by the University of North Carolina Press), 2001, 121.

34. Joseph Ball to Joseph Chinn, Joseph Ball Letterbook, 18 Feb. 1744, MS, Library of Congress.

35. Worlidge, *Vinetum Britannicum,* iii, iv, 97, 239–240.

36. Fitzhugh to George Mason, 20 July 1694, in *William Fitzhugh,* 328; William Beverley to Micajah Perry & Co., 11 July 1738 and 24 May 1739, qtd. in Ayres, "Fruit Culture," 138; "Appraisement of the Estate of Robert Babb, Dec'd, May 7, 1781," ed. R. S. Thomas, *WMQ,* ser. 1, 12, no. 1 (July 1898): 41.

37. Ayres, "Fruit Culture," 139; *Pennsylvania Gazette,* 19 May 1747.

38. Robert King Carter, Aug. 1724, "Alcohol" vertical file, Rockefeller Library; Ayres, "Fruit Culture," 145, 185; Landon Carter, "Diary," 12 Aug. 1778; Hazard, "Journal of Ebenezer Hazard," 420.

39. Robert Wormeley Carter, "Diary," 25 Jan. 1777, 6 June 1791, 17 Aug. 1791; Landon Carter, "Diary," 1:345; Toulmin, *Western Country in 1793,* 45; Henry Wynkoop, "Account of a Crab Apple Orchard, 1814," qtd. in Ayres, "Fruit Culture," 189–194.

40. Hugh Jones, *The Present State of Virginia,* ed. Richard L. Morton (1724; Chapel Hill: Published for the Virginia Historical Society by the University of North Carolina Press, 1956), 78; William Hugh Grove, "Diary 1698–1732," MS, Alderman Library, 7; Durand qtd. in Ayres, "Fruit Culture," 128–29, 33; Fitzhugh to Ralph Smith, 22 April 1686, in *William Fitzhugh.*

41. John J. McCusker, *Rum and the American Revolution: The Rum Trade and the Balance of Payments of the Thirteen Continental Colonies* (New York: Taylor and Francis, 1989), 772; Anon., *The Servants' Guide and Family Manual* (2nd ed. London, 1831), Winterthur Library, 156–57; John Worlidge, *Vinetum Britannicum,* 132–35.

42. Edwin Tunis, *Colonial Craftsmen and the Beginnings of American Industry*

(New York: World Publishing Company, 1965),121; *Pennsylvania Gazette*, 7 March 1732, 10 May 1764.

43. Robert King Carter, vertical file, Rockefeller Library; William Byrd, *The London Diary (1717–1721) and Other Writings*, ed. Louis B. Wright and Marion Tinling (New York: Oxford University Press, 1958), 505.

44. Anon., *Servants' Guide*, 183, 56, 165; Anon., *The Compleat Planter & Cyderist* (London, 1690), Winterthur Library, 217, 141; Landon Carter, "Diary," 2:671.

45. Charles Robinson to John Norton, 18 June 1772, "John Norton and Sons Papers [1763–1798]," TS, Rockefeller Library; Robert Roberts, *The House-Servant's Directory; or, a Monitor for Private Families* (2nd ed. Boston, 1828), 180; Anon., *Servant's Guide*, 52, 165, 180; *Maryland Gazette*, 11 Aug. 1747.

46. Landon Carter, "Diary," 2:1134–35; Anon., *The Maid of all-work, or General Servant: Her Duties and how to Perform Them* (London, 187?), microfilm, 183.

47. John Burnett, *Liquid Pleasures: A Social History of Drinks in Modern Britain* (London: Routledge, 1999), 161.

48. Gregory A. Stiverson and Cynthia Z. Stiverson, *Books Both Useful and Entertaining: A Study of Book Purchases and Reading Habits of Virginians in the Mid-Eighteenth Century* (Colonial Williamsburg Early American History Research Report, 1977), microfiche, Alderman, 11, 58, 60, 73–74.

49. Bennie Brown, "The Library of John Mercer of Marlborough," MS, VHS; Anon., "Title Pages of the Libraries of Landon Carter and Robert Wormeley Carter of Sabine Hall," MS, Rockefeller Library; Landon Carter, "Diary," 2:1118; Kevin J. Hayes, *The Library of William Byrd of Westover* (Madison: Madison House in cooperation with the Library Company of Philadelphia, 1997); John Evelyn, *Sylva; or, a Discourse of Forest Trees* (London, 1664), Dibner Library.

50. William Byrd to Sir Charles Wager, 12 April 1741, "Letters of the Byrd Family," *VMHB* 37, no. 2 (April 1929): 109; "Some Communications about an early swarm of Bees. Also concerning Cider; Descent of Sap; the Season of Transplanting Vegetables," in *The Philosophical Transactions of the Royal Society of London, from their Commencement, in 1665 to the year 1800; abridged, with Notes and Biographic Illustrations*, ed. Charles Hutton, George Shaw, and Richard Pearson (London, 1809), 1:580, Dibner Library; Rev. Henry Miles, "On Some Improvements which may be made in Cider and Perry," *Philosophical Transactions*, 9:165–66; Paul Dudley, "An Account of a new sort of Molasses, made of Apples," *Philosophical Transactions*, 6: 618–19.

51. "Appraisement of James Nimmo, 1754," *The Lower Norfolk County Virginia Antiquary*, ed. Edward W. James (1895; New York: reprinted by Peter Smith, 1951), 1:91, n. 3; *Virginia Gazette*, Rind, 12 March 1767; "Inventory of Estate of John Andrews Sept. 21, 1719," York County Wills and Inventories, 1716–1720, TS, Rockefeller Library, 15:490–91; "Inventory of Estate of William Anthony, Jan. 18, 1731," York County Wills & Inventories 1729–1732, TS, Rockefeller Library, 17:143; Jean Russo at the Maryland Historic Society graciously provided the counts on stills in Southampton, St. Mary's, York, Talbot, Somerset, Ann Arundel, and Kent counties.

52. "Will of William Nimmo Sr. of Prince Anne County, 17 Jan. 1791," *Lower Norfolk County Virginia Antiquary*, 1:93; Headley, *Wills of Richmond County*, 112, 90, 126.

53. *Virginia Gazette*, Rind, 16 June 1738; James Carter to John Norton 29 May 1769, "John Norton & Sons papers [1763–1798]," TS, Rockefeller Library; Peter Lyons to John Norton, 24 Sept. 1768, *John Norton and Sons Papers*; *Virginia Gazette*, 11 July, 1766; Landon Carter, *Diary*, 2:1149.

54. George Smith, *The Practical Distiller, or a Brief Treatise on Practical Distillation* (London, 1718), 29, LCP; Israel Pemberton, *Letterbook 1744–1747*, 23 June 1745, ms., APS, 67; "Directions from Mr. Samuel Burge for the Distillation of Spirits from Grain," in Correspondence of Baynton and Morgan 1766, microfilm, David Library.

55. *Virginia Gazette*, Rind, 12 March 1767.

56. Joseph Ball to Joseph Chinn, 25 March 1742/43, Joseph Ball "Letterbook, 1734–1780," MS, LC; Ayres, "Fruit Culture," 142. For price, see Robert Wormeley Carter, "Diary," 5 Aug. 1765, 3 Dec. 1765, 11 Oct. 1774, 15 Oct. 1781.

57. Robert Rose, *The Diary of Robert Rose: A View of Virginia by a Scottish Colonial Parson 1746–1751*, ed. Ralph Emmett Fall, 23 Jan. 1748, (Verona, VA: McClure Press, 1977), 49, 245, n. 516.

CHAPTER FOUR: "ANNE HOWARD . . . WILL TAKE IN GENTLEMEN"

1. Ruth H. Bloch, "The Gendered Meanings of Virtue in Revolutionary America," *Signs: Journal of Women in Culture and Society* 13, no. 1 (1987): 51, 46; *Virginia Gazette*, 4 March 1773.

2. David W. Conroy, *In Public Houses: Drink and the Revolution of Authority in Colonial Massachusetts* (Chapel Hill: Published for the OIEAHC by the University of North Carolina Press, 1995); Peter Thompson, *Rum Punch and Revolution: Taverngoing and Public Life in Eighteenth-Century Philadelphia* (Philadelphia: University of Pennsylvania Press, 1999), 25, 51, 5.

3. *Virginia Gazette*, 22 December 1768; Kym Rice, *Early American Taverns: For the Entertainment of Friends and Strangers* (Chicago: Regnery Gateway, 1983), 91; "The Fisher History," *Some Prominent Virginia Families*, ed. Louise P. duBellet (Lynchburg, VA, 1907), 2: 788.

4. *Virginia Gazette*, 4 March 1773; Patricia Gibbs, "Taverns in Tidewater Virginia, 1700–1774" (master's thesis, College of William and Mary, 1968), 203; *Virginia Gazette*, 11 April 1751, 20 March 1752; "Virginia Gazette Day Book, 1750–1752," MS, Alderman Library, 31; Gibbs, "Taverns," 31; Lindsay O. Duvall, *York County Wills and Inventories XVIII* (Easley, SC: Southern Historical Press, 1982), 57; Robert Wormeley Carter, "Diary 1765–1792," 16 Dec. 1756, TS, Rockefeller Library; *Maryland Gazette*, 26 Nov. 1763; Salinger, *Taverns and Drinking*, 74–75; *Virginia Chronicle and Norfolk and Portsmouth General Advertiser*, 26 Jan. 1793, microfilm, David Library.

5. Robert Wormeley Carter, "Diary," 6 August 1790; Peter Legaux, "Journal of the Pennsylvania Vine Company," 20 March 1804, MS, APS; Dr. Alexander Hamilton, *The Tuesday Club: A Shorter Edition of The History of the Ancient and Honorable*

Tuesday Club, ed. Robert Micklus (Baltimore: Published for the IEAHC by Johns Hopkins University Press, 1995); G. Kurt Piehler, "Phi Beta Kappa: The Invention of an Academic Tradition," *History of Education Quarterly* 28, no. 2 (1988): 218.

6. *Virginia Gazette*, 11 April 1751.

7. Gibbs, "Taverns," 99; *Virginia Gazette*, 8 May 1770.

8. Edmund S. Morgan, *American Slavery: American Freedom: The Ordeal of Colonial Virginia* (New York: W. W. Norton, 1975), 151; *Virginia Gazette*, 1 Aug. 1766, 25 July 1771; Byrd, *Secret Diary*, 7 Nov. 1711; Anne Coke Pattison, "Account Book, 1743/4 January 7–1749 June 13," 9 Aug. 1744, MS, VHS; Warville qtd. in Rice, *Early American Taverns*, 67.

9. David John Mays, *Edmund Pendleton 1721–1803* (1952; Richmond: Virginia State Library, 1984), 1:29; Ferdinand Marie Bayard, *Travels of a Frenchman in Maryland and Virginia*, ed. and trans. Ben C. McCary (1791; Williamsburg, VA: Edwards Brothers, 1950), 11; Rice, *Early American Taverns*, 64; Julia Cherry Spruill, *Women's Life and Work in the Southern Colonies* (New York: W.W. Norton, reprint 1972), 301.

10. Weynette Parks Haun, *Surry County, Virginia Court Records* (Durham, NC: private publication, 1987); A. W. Bohanan, *Old Surry: Thumb-nail Sketches of Places of Historic Interest in Surry County, Virginia* (Petersburg, VA: Plummer Printing, 1927), 16; Mary A. Stephenson, *Old Homes in Surry and Sussex* (Richmond, VA: Dietz Press, 1942), 25; James D. Kornwolf, *So Good a Design: The Colonial Campus of the College of William and Mary* (Williamsburg, VA: College of William and Mary, 1989), 10.

11. Stephenson, *Old Homes*, 25; Kornwolf, *So Good a Design*, 105. John Ruffin was a burgess from 1738–41, 1744–47, and in 1754.

12. Kornwolf, *So Good a Design*, 99–100; Haun, *Surry Court Records*, 49, 77; A. G. Roeber, *Faithful Magistrates and Republican Lawyers: Creators of Virginia Legal Culture, 1680–1810* (Chapel Hill: University of North Carolina Press, 1981), 46.

13. Stephenson, *Old Homes*, 66–67.

14. *Virginia Gazette*, 4 April 1748; 10 Dec. 1773; March 1768; 20 June 1766; 3 Oct. 1771.

15. Gibbs, "Taverns," 137, n. 12; Wallace B. Gusler, "Anthony Hay: A Williamsburg Tradesman," in *Common People and Their Material World: Free Men and Women in the Chesapeake, 1700–1830: Proceedings of the March 13, 1992 Conference*, ed. David Harvey and Gregory J. Brown (Colonial Williamsburg Research Publications, 1995), 24–26; Mathew Moody's cabinetmaking, *Virginia Gazette*, 25 April 1766; Rust and Markland in Roeber, *Faithful Magistrates*, 87 and 95; Brown, *Good Wives*, 459 n. 31; *Maryland Gazette*, 25 Aug. 1757; Gibbs, "Taverns," 153–54; 198–99.

16. Gibbs, "Taverns," 158, 198–99; Duvall, *York County Wills*, 15:251, 3:440–41; Gibbs, "Taverns," 172; Kierner, *Beyond the Household*, 20, 24.

17. Francois Jean Marquis de Chastellux, *Travels in North America, in the Years 1780, 1781, and 1782: A Revised Translation*, ed. Howard C. Rice Jr. (Chapel Hill: Published for the IEAHC by the University of North Carolina Press, 1963), 2:404; "Will of John Bramham, 28 March 1754," *Wills of Richmond County, Virginia, 1699–1800*, ed. Robert K. Headley Jr., (Baltimore: Clearfield, 1993), 113; William Byrd, "The Secret

History of the Line," in *Histories of the Dividing Line betwixt Virginia and North Carolina* (1929; New York: Dover , 1967), 37; Francisco de Miranda, *The New Democracy in America: Travels of Francisco de Miranda in the United States, 1783–1784*, ed. John S. Ezell, trans. Judson P. Wood (Norman: University of Oklahoma Press, 1963), 10; Bierce, *Travels in the Southland*, 49; Sturtz, *Within Her Power*, 107.

18. Duvall, *York County Wills*, 17:77, 184; 15:562.

19. Gibbs, "Taverns," 200.

20. "Will of Governor Fauquier," in Duvall, *York County Wills*, 22:95–99; *Virginia Gazette*, 6 Oct. 1768; Gibbs, "Taverns," 148; William Walter Hening, *The Statutes at Large: Being a Collection of all the Laws of Virginia, from the First Session of the Legislature in the year 1619* (New York: 1819–1823), April 1691, 2:45; Oct. 1705, 3:396; Aug. 1734, 4:428.

21. Hening, *Statutes at Large*, Aug. 1734, 4:428.

22. Hening, *Statutes at Large*, Oct. 1644, 1:287; March 1646, 319; Nov. 1647, 350; March 1658, 446.

23. Hening, *Statutes at Large*, Nov. 1762, 7: 595.

24. Peter Kalm, *Travels in North America, 1748–1749*, trans. John Reinhold Forester (Barre, MA: Imprint Society, 1972), 204; Byrd, *Secret Diary*, 12 April 1712; Duvall, *York County Wills*, 2:268.

25. Roeber, "Authority, Law, and Custom," 46.

26. Hening, *Statutes at Large*, Oct. 1705, 3: 396–401; Alice Hanson Jones, *Wealth of a Nation to Be: The American Colonies on the Eve of the Revolution* (New York: Columbia University Press, 1980), 63. £11.2 was sterling.

27. Charles Carter (presumed), "Ship Tavern Ledger 1752–1758," MS, VHS.; Jerrilynn Eby, *They Called Stafford Home: The Development of Stafford County, Virginia from 1600 until 1865* (Bowie, MD: Heritage Books, 1997), 332–35.

28. Presumably the tavern was sold or went under, because by 1762 Charles Carter of Cleve was very ill and not expected to live. Charles Carter Jr. of Cleve to Landon Carter, Landon Carter Papers, 27 July 1762, MSS, Alderman Library. Landon Carter advertised in the *Virginia Gazette* that he was selling one hundred slaves who had belonged to his deceased son on 25 June 1767; Account of William Abbot, "Ship Tavern Ledger," 212.

29. On July 11, 1766, Landon Carter and Charles Carter Jr. advertised in the *Virginia Gazette* that they were selling Charles Carter of Cleve's eight copper stills. He also owned a 500-gallon copper and a 300-gallon copper. Since it is unclear whether the stills and coppers were for Charles Carter of Cleve's plantations or the tavern, the cost of these items was not included in the profit analysis.

30. Salinger, *Taverns and Drinking*, 107, 104; *Maryland Gazette*, 3 Aug. 1748, 10 Feb. 1749; *Virginia Gazette*, 12 Sept. 1771; *Maryland Gazette*, 27 Oct. 1768; Roeber, "Authority, Law, and Custom," 42; Bierce, *Travels in the Southland*, 77.

31. Stephenson, *Old Homes*, 26. £60 is from 1747, tobacco from 1748.

32. Kierner, *Beyond the Household*, 20.

33. Rice, *Early American Taverns*, 31, 47; Alexander Macaulay, "Journal of Alexan-

der Macaulay," *WMQ* 11, no. 3 (Jan. 1903): 186; Landon Carter, "Diary," 2:623; Nicholas Cresswell, *The Journal of Nicholas Cresswell 1774–1777*, ed. Lincoln Mac-Veagh (New York: Dial Press, 1924), 124.

34. Byrd, *Secret Diary*, 488; Heather R. Wainwright, "Inns and Outs: Anne Patti-son's Tavern Account Book, 1744–1749" (master's thesis, Armstrong Atlantic State University, 1998), 96; Salinger, *Taverns and Drinking*, 90, 113; Cresswell, *Journal*, 20.

35. Wainwright, "Inns and Outs," 105.

CHAPTER FIVE: "LADYS HERE ALL GO TO MARKET TO SUPPLY THEIR PANTRY"

1. Mary Goodwin, "Markets and Fairs," *Colonial Williamsburg Research Reports* (1950), microfiche, Alderman Library; Gregory J. Brown, "Distributing Meat and Fish in Eighteenth-Century Virginia: The Documentary Evidence for the Existence of Markets in Early Tidewater Towns," *Colonial Williamsburg Research Reports* (1988), microfiche, Rockefeller Library, 12.

2. Brown, "Distributing Meat and Fish," 12–13; Goodwin, "Markets and Fairs," 12–18; Walsh et al., *Provisioning Early American Towns*, 91.

3. Goodwin, "Markets and Fairs," 6; James H. Soltow, *The Economic Role of Williamsburg* (Williamsburg: Colonial Williamsburg, 1965), 10; *Virginia Gazette*, 29 June 1769.

4. Soltow, *Economic Role of Williamsburg*, 16–18.

5. Ibid., 26–27; Jacob Price, *Tobacco in Atlantic Trade: The Chesapeake, London, and Glasgow, 1675–1775* (Brookfield, VT: Variorum, 1995).

6. Soltow, *Economic Role of Williamsburg*, 42.

7. Robert Polk Thomson, "The Merchant in Virginia 1700–1775" (Ph.D. diss., University of Wisconsin, 1955), 115, 186, 10; Soltow, *Economic Role of Williamsburg*, 48; John Hook, "Letterbook," MS, LVA; Ann Smart Martin, "Buying into the World of Goods: Eighteenth-Century Consumerism and the Retail Trade from London to the Virginia Frontier" (Ph.D. diss., College of William & Mary, 1993), 180.

8. Thomson, "Merchant in Virginia," 184; Martin, "Buying," 202; Anon., "A Letter Written in Caroline County, Portobago Bay on the Rappahannock River, March 24, 1785," *VMHB* 23, no. 4 (Oct. 1915): 412.

9. Ivor Noel Hume, "Beverage Bottles in Colonial Virginia: A Survey Based on Discoveries from the Vicinity of Colonial Williamsburg," *Colonial Williamsburg Research Reports* (1961), microfiche, Alderman Library.

10. Daniel Roberdeau to John Thornton, 9 Sept. 1774, "Daniel Roberdeau Letterbook, 1767–1791," MS, LC; William Allason to Robert Allason, 12 Sept. 1757, and William Allason to Captain Walker, 8 Nov. 1757, "Letterbook of William Allason, 1757–1770," MS, LVA.

11. Martin, "Buying," 305, 302.

12. Ibid.

13. *Alexandria Gazette*, 3 Aug. 1803, 4 Aug. 1803, 5 Aug. 1803.

14. *Virginia Gazette*, 8 Dec. 1777, 16 Aug. 1776, 8 May 1778, 4 Nov. 1773, 25 Sept. 1775, 5 March 1767.

15. John J. McCusker, *Rum and the American Revolution: The Rum Trade and the Balance of Payments of the Thirteen Continental Colonies* (New York: Garland, 1989), 411–44, 468–79; John Brewer and Roy Porter, *Consumption and the World of Goods* (New York: Routledge, 1994), 183.

16. Anon., *The Compleat Planter and Cyderist* (London, 1690), vii, Winterthur Library; George Watkins, *The Compleat English Brewer* (London, 1768), vii, 66, Winterthur Library; Harrison Hall, *Hall's Distiller* (Philadelphia, 1813), 137, Dibner Library.

17. Col. William Fitzhugh to Capt. Francis Partis, 11 June 1680, "Letterbook," William Fitzhugh Papers, photostat, VHS, 17–18; Robert Beverly to Landon Carter, 16 Jan. 1766, Landon Carter Papers, 1763–1774, MSS, VHS; John Baylor to Messrs. Cary, 6 Sept. 1760, "John Baylor Letterbooks, 1749–1753, 1757–1765," photostat, LVA; Richard Corbin to Charles Govre, 15 June 1758, "Letterbook 1758–1768," photostat, VHS; John Custis, 12 Aug. 1724, "John Custis Letterbook 1717–1744," TS, Rockefeller Library.

18. Col. Francis Taylor, "Diary," 3 July 1786, microfilm, LVA; George Washington to Landon Carter, 18 June 1786, in *The Papers of George Washington*, ed. Dorothy Twohig (Charlottesville: University Press of Virginia, 1998-); Baylor to Messrs. Cary, 6 Sept. 1760, "John Baylor Letterbook"; Custis to unknown recipient, 12 Aug. 1724, "John Custis Letterbook."

19. T. H. Breen, *Tobacco Culture: The Mentality of the Great Tidewater Planters on the Eve of the Revolution* (Princeton: Princeton University Press, 1985).

20. Lorena S. Walsh, "Provisioning Tidewater Towns," *Explorations in Early American Culture* 4 (2000): 66; Francis Jerdone to Neil Buchanan, 29 May 1741, Jerdone Papers, MS, Swem Library.

21. John Fitzgerald to George Washington, 12 June 1797, *Papers of George Washington*, 181; John Taylor, *Arator: Being a Series of Agricultural Essays, Practical and Political in Sixty-Four Numbers*, ed. M. E. Bradford (1818; Indianapolis: Liberty Fund, 1977), 272, VHS. Taylor's essays first appeared in 1803 in a Georgetown newspaper and were published in book form in 1813.

22. Philip D. Morgan, *Slave Counterpoint: Black Culture in the Eighteenth Century Chesapeake and Low Country* (Chapel Hill: Published for the OIEAHC by the University of North Carolina Press, 1998), 39; George Washington, *Weekly Farm Reports 1786–1797*; Ann Lucas, "Spring Dinner."

23. "Will of William Dudley, 21 Dec. 1751," in *Wills of Richmond County, Virginia, 1699–1800*, ed. Robert K. Headley Jr. (Baltimore: Clearfield Company, 1993), 107.

24. Helen Hill Miller, "A Portrait of an Irascible Gentleman: John Mercer of Marlborough, *Virginia Cavalcade* 26:2 (Autumn, 1976): 75, 78.

25. Qtd. in Malcolm C. Watkins, *The Cultural History of Marlborough* (Washington, DC: Smithsonian Institution Press, 1968), 55.

26. John Mercer to George Mercer, *George Mercer Papers: Relating to the Ohio Company of Virginia*, ed. Lois Mulkearn (Pittsburgh: University of Pittsburgh Press, 1954), 193.

27. George Harrison Sanford King, "Notes from the Journal of John Mercer, esquire of Marlborough, Stafford County, Virginia," MS, VHS; *Virginia Gazette*, 18

April 1766; John Mercer to George Mercer, 22 Dec. 1767 to 28 Jan. 1768, in *George Mercer Papers*, 186–220; D'Anmours to Thomas Jefferson, 27 Feb. 1782, in *The Papers of Thomas Jefferson*, vol. 6, ed. Julian P. Boyd (Princeton: Princeton University Press, 1950).

28. *George Mercer Papers*, 194; "John Mercer Account Book 1731–1767," MS, VHS; Will of John Mercer, December 12, 1768, and Will of Ann Mercer, 8 Oct. 1770, photostat, VHS.

29. George Washington to Lund Washington, 26 Nov. 1775, *Papers of George Washington*, 1:11.

30. George Washington to John Fitzgerald, 12 June 1796, *Papers of George Washington*, 180–81; James Anderson to George Washington, 21 June 1797, *Papers of George Washington*, 199.

31. George Washington to Henry Lee, 25 Jan. 1798, *Papers of George Washington*, 46; George Washington to Robert Lewis, 26 Jan. 1798, *Papers of George Washington*, 47; George Washington to James Anderson, 22 May 1798; George Washington to James Anderson, 6–7 Feb. 1798, *Papers of George Washington*, 74–75; www.mount vernon.org/visit/plan/index.cfm/pid/807/.

32. Daniel Roberdeau, "Letterbook 1767–1791," MS, LC, 123, 136, 184, 95.

33. Ibid., 183, 216.

CHAPTER SIX: "EVERY MAN HIS OWN DISTILLER"

1. Londa Schiebinger, *The Mind Has No Sex? Women in the Origins of Modern Science* (Cambridge: Harvard University Press, 1989), and *Nature's Body: Gender in the Making of Modern Science* (Boston: Beacon Press, 1993); *Virginia Gazette*, Purdie and Dixon, 5 Aug. 1773; "On the Application of Chemistry to Agriculture, and the Rural Economy," *Columbian Magazine* 2 (1788): 754–57.

2. Eliza Smith, *The Compleat Housewife* (1727; London: Studio Editions Ltd, 1994), 254; Alexander Morrice, *A Treatise on Brewing* (London, 1802), 30, 53, 175, Dibner Library; George Watkins, *The Compleat English Brewer* (London, 1768), 50, Winterthur Library.

3. John Tuck, *The Private Brewer's Guide to the Art of Brewing* (1817; London, 1822), 181, Dibner Library; Thomas Thomson, *Brewing and Distillation* (Edinburgh, 1849), 9–10, Dibner Library; Landon Carter, *Diary*, 2:697.

4. William Irvine, *Essays, Chiefly on Chemical Subjects* (London, 1805), microfilm, Alderman Library; Thomson, *Brewing and Distillation*, 77, 72; Harrison Hall, *Hall's Distiller* (Philadelphia, 1813), 7, Dibner Library.

5. Morrice, *Treatise on Brewing*, vii; Hall, *Hall's Distiller*, iii; Tuck, *Private Brewer's Guide*, title page, v, viii.

6. Robert May, *The Accomplisht Cook, or the Art and Mystery of Cookery*, ed. Alan Davidson, Marcus Bell, and Tom Jaine (1660, 1685; Devon, England: Prospect Books, 1994), 28; Richard Bradley, *The Country Housewife and Lady's Director*, ed. Caroline

Davidson (1727; London: Prospect Books, 1980), 51–52; Morrice, *Treatise on Brewing,* vii, 34, 45; Tuck, *Private Brewer's Guide,* 244; Thomson, *Brewing and Distillation,* 4.

7. Morrice, *Treatise on Brewing,* title page and preface (n.p.).

8. Ibid., preface and second preface, n.p.

9. Raymond Phineas Stearns, *Science in the British Colonies of America* (Urbana: University of Illinois Press, 1970), 280; Ayres, "Fruit Culture," 71, 72, 75; Brooke Hindle, *The Pursuit of Science in Revolutionary America, 1735–1789* (Chapel Hill: Published for the IEAHC by the University of North Carolina Press, 1956), 28, 158.

10. *Journal and Letters of Philip Vickers Fithian, 1773–1774: A Plantation Tutor of the Old Dominion,* ed. Hunter Dickinson Farish (Princeton: Princeton University Press for Colonial Williamsburg, 1943), 116; Ayres, "Fruit Culture," 17.

11. Landon Carter, "Diary," 2:697, 931, 932, 1118, 1135, 1149, 1:192, 1:534; "Inventory of Estate of Landon Carter Esq., Feb. 1779," MS, VHS; *Virginia Gazette,* Purdie, 17 Feb. 1775, supplement.

12. Digbie, *The Closet,* xxx; Bradley, *Country Housewife,* vii–viii, 39, 51–52; Martha Washington, *Martha Washington's Booke of Cookery and Booke of Sweetmeats,* transcribed by Karen Hess (New York: Columbia University Press, 1981), 392, 381.

13. Jane Carson, *Colonial Virginia Cookery: Procedures, Equipment, and Ingredients in Colonial Cooking* (Williamsburg: Colonial Williamsburg Foundation, 1985), xv; Hannah Glasse, *The Art of Cookery, Made Plain and Easy* (London, 1747), LCP; William Ellis, *The Country Housewife's Family Companion,* 6th ed. (London, 1750), LCP.

14. Martha Bradley, *The British Housewife* (London, 1770), 85, microfilm, Alderman Library; Susannah Carter, *The Frugal Colonial Housewife,* ed. Jean McKibbin (Garden City, NY: Dolphin Books, 1976); Mary Cole, *The Lady's Complete Guide* (London, 1791), 361, Winterthur Library.

15. Amelia Simmons, *The First American Cookbook: A Facsimile of "American Cookery,"* ed. Mary Tolford Wilson (1796; New York: Dover, 1958), xvii, xviii, xix, 47.

16. William Ellis, *The Practical Farmer* (London, 1732, LCP; George Cooke, *The Complete English Farmer* (London, 1741), LCP; George Smith, *A Compleat Body of Distilling,* 3rd ed. (London, 1738), Dibner Library; John Edgar Molnar, "Publication and Retail Book Advertisements in the Virginia Gazette, 1736–1780" (Ph.D. diss., University of Michigan, 1978), 246; Ambrose Cooper, *The Complete Distiller* (London, 1757), vi, 75, Dibner Library.

17. Thomas Chapman, *The Cyder-Maker's Instructor* (reprint. Philadelphia, 1760), title page, iv, LCP; Watkins, *Compleat English Brewer,* vii; Elijah Bemiss, *The Dyer's Companion: In Two Parts* (New York, 1806), 299, 302, LCP.

18. Morrice, *Treatise on Brewing,* 19.

19. "Will of Bartholomew Andrews, 28 Nov. 1720," *Wills and Administrations of Surry County, Virginia, 1671–1750,* ed. Eliza Timberlake Davis (Baltimore: Clearfield, 1995), 5; "Will of William Walker, 15 April 1750," *Wills of Richmond County, Virginia, 1699–1800,* ed. Robert K. Headley Jr. (Baltimore: Clearfield, 1993), 112; Headley, 17; June Whitehurst Johnson, *Fairfax County Virginia Will Book A 1742–1752, Will*

Book B 1752–1767 (Berryville, VA: Virginia Book Co., 1982), 52; Headley, 122, 134; "Will of Thomas Haynes, 1746," Virginia Colonial Records Project, Report no. 4865, LVA (online); "Joseph Carter, Spotsylvania, Will Feb. 19, 1750," *Carter Genealogy*, Joseph Lyon Miller, *WMQ* 18, no. 1 (July 1909), 53.

20. Morrice, *Treatise on Brewing*, 175; Tuck, *Private Brewer's Guide*, 17.

21. The first known side distillation apparatus was described by Filippo Ulsted of Nurember in 1526. Its invention date is not known.

22. Smith, *A Compleat Body of Distilling*; A. Cooper, *The Complete Distiller* (London 1757), 52, Winterthur.

23. Harrison Hall, *Hall's Distiller* (Philadelphia, 1813), 5–6, 10–11, Dibner Library.

24. Michael Krafft, *The American Distiller* (Philadelphia, 1804), introduction, Dibner Library.

25. *Virginia Gazette*, 26 Sept. 1766, 8 March 1769, 5 Oct. 1769, 6 Feb. 1772; Ayres, "Fruit Culture," 137; 144–45.

26. David C. Klingaman, "The Development of Virginia's Coastwise Trade and Grain Trade in the Late Colonial Period" (Ph.D. diss., University of Virginia, 1967), 70, 87, 90, 92, 88.

27. Molnar, "Publication and Retail Book Advertisements," 240, 244, 204, 199.

28. Molnar, "Publication and Retail Book Advertisements," 190; T. T. Philomath, *The Virginia Almanack for the Year of Our Lord God* 1770 (Williamsburg, 1769); *The American Farmer*, ed. John S. Skinner, 3rd ed. (Baltimore, 1821), see, 2 and 16 April 1819 for persimmon beer; 25 June 1819 for ginger beer; 13 Aug. 1819, 15 Oct. 1819, 5 Nov. 1819, and 11 Feb. 1820 for cider; 21 Jan. 1820 for brown spruce beer; 4 Feb 1820 for white spruce beer; and 11 Feb. 1820 for fruit trees, VHS.

29. Barbara McEwan, *Thomas Jefferson: Farmer* (Jefferson, NC: McFarland, 1991), 76–80; *Virginia Gazette*, 17 Feb. 1775; Peter Legaux, "Journals of the Pennsylvania Vine Company," 2:41, MSS, APS; *Mercer Family Papers*, n.d., mss., VHS; *American Farmer*, 23 April 1819.

30. John Worlidge, *The Second Part of Vinetum Britannicum; or, a Treatise of Cider* (London, 1689), Winterthur Library. The cost is given in John Worlidge, *Mr. Worlidge's Two Treatises* (London, 1694), 124, Dibner Library; Ayres, "Fruit Culture," 134, 138.

31. Worlidge, *Two Treatises*, 128; Ayres, "Fruit Culture," 142.

32. Ayres, "Fruit Culture," 34; Peter J. Hatch, *The Gardens of Monticello* (Charlottesville: Thomas Jefferson Memorial Foundation, 1992), 50; William Coxe, *A View of the Cultivation of Fruit Trees* (Philadelphia, 1817), 78; *Virginia Gazette*, 5 Nov. 1772; S. W. Fletcher, *History of Fruit Growing in Virginia* (Staunton, VA: n.p., 1932), 3–4.

33. Ayres, "Fruit Culture," 132, 133; author's discussion with Peter Hatch, Head of Gardens and Orchards, Monticello, 30 Jan. 2002; *Virginia Gazette*, 17 Feb. 1762, 26 Sept. 1755; Robert Wormeley Carter, "Diary," 17 July 1780, TS, Rockefeller Library; Jerdone Family Papers 1762–1801, 8 Nov. 1790, MSS, Swem Library.

34. Lois Green Carr, "Diversification in the Colonial Chesapeake: Somerset County, Maryland, in Comparative Perspective," in *Colonial Chesapeake Society*, ed. Lois Green Carr and Philip D. Morgan (Chapel Hill: University of North Carolina

Press, 1991), 372, 376; Francis Louis Michel, "Report of the Journey of Francis Louis Michel from Berne, Switzerland, to Virginia October 2, 1701 — December 1, 1702," trans. and ed. William J. Hinke, *VMHB*, 24 no. 2 (April 1916): 18, 19; Ferdinand Marie Bayard, *Travels of a Frenchman in Maryland and Virginia, 1791*, trans. and ed. Ben C. McCary (1791; Williamsburg, VA: Edwards Brothers, 1950), 58, 81; J. P. Brissot de Warville, *New Travels in the United States of America*, ed. Durand Echeverria (Cambridge: Belknap Press of Harvard University Press, 1964), 341; Ayres, "Fruit Culture," 138, 139; "A Letter Written in Caroline County," 410; qtd. in Ayres, "Fruit Culture," 131.

35. Erma Risch, *Quartermaster Support of the Army: A History of the Corps 1775–1939* (Washington, DC: Quartermaster Historian's Office, Office of the Quartermaster General, 1962), 57; Charles Royster, *A Revolutionary People at War: The Continental Army and American Character, 1775–1783* (Chapel Hill: Published for the IEAHC by the University of North Carolina Press, 1979), 191; Theodore Thayer, *Nathaniel Greene: Strategist of the American Revolution* (New York: Twayne Publishers, 1960), 93; Erma Risch, *Supplying Washington's Army* (Washington, DC: Center of Military History United States Army, 1981), 193–94.

36. Knox qtd. in Royster, *Revolutionary People*, 75; George Washington to Robert Morris, 27 Sept. 1781, in *The Papers of Robert Morris*, ed. E. James Ferguson (Pittsburgh: University of Pittsburgh Press, 1975), 362–63; Holly A. Mayer, *Belonging to the Army: Camp Followers and Community during the American Revolution* (Columbia: University of South Carolina Press, 1996), 141.

37. Royster, *Revolutionary People*, 249.

38. Ibid., 144.

39. Risch, *Supplying Washington's Army*, 190; *Journal of the Continental Congress 1774–1789* (Washington, DC, 1904–1937), 3:322, microfiche, Alderman Library; Carp, *To Starve the Army at Pleasure*, 24; Bennett, *Ale, Beer, and Brewsters*, 93–94; Richard Backhouse, *Account Book*, 1775, 27 July 1775, ms., HSP; Juliet Fauntleroy, "Revolutionary Claims of Campbell County, Virginia, March 1782 to Aug. 1786," 1:15, 20, 27, MS, LVA; Captain Roger, Maryland State Paper Series D, MS, Hall of Records.

40. Walter Hart Blumenthal, *Women Camp Followers of the American Revolution* (Philadelphia: George S. MacManus, 1952), 49; Mayer, *Belonging to the Army*, 122.

41. Fred Anderson, *A People's Army: Massachusetts Soldiers and Society in the Seven Years' War* (Chapel Hill: Published for the IEAHC by the University of North Carolina Press, 1984), 127; Mayer, *Belonging to the Army*, 225.

42. Chris Leffingwill to Joseph Trumbull, 7 April 1776, Papers of Joseph Trumbull 1753–1791, microfiche, David Library; Theodore Mumford to Joseph Trumbull, 27 June 1776, Papers of Joseph Trumbull; Risch, *Supplying Washington's Army*, 192, 206.

43. Risch, *Supplying Washington's Army*, 190; See Rorabaugh, *Alcoholic Republic*, 67–73, on the rise of whiskey; Carp, *To Starve the Army*, 21; Blumenthal, *Women Camp Followers*, 23, 46–47.

44. For example, "Advertisement for West Point Contract Proposals," *Papers of Robert Morris*, 22 Oct. 1781, 3:100; Morris, "Diary," 21 Aug. 1781, *Papers of Robert Morris*, 2:77; Morris to Ridley and Pringle, 19 Oct. 1781, "Diary," 2:96.

45. Mayer, *Belonging to the Army*, 94, 7–10; Anderson, *A People's Army*, 119; Mayer, *Belonging*, 14, 49–50, 243–44.

46. Colonel Francis Taylor, "Diary 1786–1799," microfilm, LVA. Taylor's diary for 1793 has been lost. Ann L. Miller, *Antebellum Orange: The Pre-Civil War Homes, Public Buildings and Historic Sites of Orange County, Virginia* (Orange, VA: Moss Publications, 1988), 113–14, 121; Barbara Vines Little, *Orange County, Virginia, Tithables 1734–1782. Part One* (Orange, VA: Dominion Market Research Corporation, 1988), 132.

47. Taylor, "Diary." Joseph Ball to Joseph Chinn, 18 Feb. 1744, "Joseph Ball Letterbook 1743–1780," MS, LC, 118.

48. Taylor, "Diary," 19 April 1789, 19 Sept. 1789, 17 March 1788.

49. Taylor, "Diary," 14 Oct. 1786, 25 Aug. 1792, 1 Nov. 1792, 24 Sept. 1796, 28 Dec. 1786, 8 Feb. 1790.

50. Taylor, "Diary," 8 Feb. 1790, 28 Oct. 1786, 28 Sept. 1790, 10 Sept., 1792, 25 May, 1792.

51. Taylor, "Diary," 17 April 1787, 20 July 1789, 6 Nov. 1789, 8 March 1790, 10 May 1788.

52. Taylor, "Diary," 6 March 1788, 22 May 1792.

53. Sarah Fouace Nourse, "Diary 1781–1783," 8 May 1781, 31 Oct. 1781, 10 Nov. 1781, 9 Nov. 1781, 7 Dec. 1781, TS, Rockefeller Library.

54. Colonel James Gordon, "Journal 1758–1768," April 1758, 10 July 1758, 3 Sept. 1758, March 1758, photostat, VHS.

CHAPTER SEVEN: "HE IS MUCH ADDICTED TO STRONG DRINKE"

1. "Northampton Order Book 4, October 1658 to June 28, 1661," ed. Susie May Ames, 30 April 1660, MS, VHS.

2. Gervase Markham, *Country Contentments, or The English Huswife* (London, 1623), 16 microfilm, Alderman Library; George Savile, *The Lady's New-Year's Gift; or Advice to a Daughter* (London, 1688), 38–43, microfilm, Alderman Library.

3. C. C. Pearson and J. Edwin Hendricks, *Liquor and Anti-Liquor in Virginia, 1619–1919* (Durham, NC: Duke University Press, 1696); Hening, *Statutes at Large*, 6 Feb. 1632, 1:167.

4. "Northampton County Virginia Court Order Book, 28 March 1654 to 28 Jan. 1655," ed. Susie May Ames, 29 March 1654, MS, VHS; "Northampton Order Book 4, October 1658 to June 28, 1661," ed. Susie May Ames, 30 July 1660, 4:114; McKey, *Accomack County*, March 1694, 8:142; *Virginia Colonial Abstracts: Northumberland County, Virginia 1678–1713*, ed. Lindsay O. Duvall, ser. 2 (Easley, SC: Southern Historical Press, 1979), 1:52. Normally cases involving servants and alcohol did not go to court, but this was a complicated case in which the servant later sold the rum to two men who fought over it; *Court Order Book One, Amelia County, Virginia, 1735–1746*, ed. Gibson Jefferson McConnaughey, 14 May 1736, (Amelia, VA: Mid-South Publishing Company, 1985).

5. Hening, *Statutes at Large*, 2:384.

6. Hening, *Statutes at Large*.

7. *Maryland Gazette*,16 Sept. 1745, 21 June 1745; Landon Carter, "Diary," 1:363, 347; Byrd, *Secret Diary*, 425, 118.

8. Benjamin Rush, *Inquiry into the Effects of Spirituous Liquors on the Human Body and Mind*, 2nd ed., microcard, Alderman Library.

9. See W. J. Rorabaugh, *Alcoholic Republic, an American Tradition* (New York: Oxford University Press, 1979).

10. Joseph R. Gusfield, *Symbolic Crusade: Status Politics and the American Temperance Movement* (Urbana: University of Illinois Press, 1986); Norman H. Clark, *Deliver Us From Evil: An Interpretation of American Prohibition* (New York: W.W. Norton, 1976); Paul E. Johnson, *A Shopkeeper's Millennium: Society and Revivals in Rochester, New York, 1815–1837* (New York: Hill and Wang, 1978); Mary P. Ryan, *Cradle of the Middle Class: The Family in Oneida County, New York, 1790–1865* (New York: Cambridge University Press, 1981); Thomas R. Pegram, *Battling Demon Rum: The Struggle for a Dry America, 1800–1933* (Chicago: Ivan R. Dee, 1998), 15–18.

11. Joe L. Coker, *Liquor in the Land of the Lost Cause: Southern White Evangelicals and the Prohibition Movement* (Lexington: University Press of Kentucky, 2007), 19–25.

12. Wolfgang Schivelbusch, *Tastes of Paradise: A Social History of Spices, Stimulants, and Intoxicants* (New York: Pantheon 1992), 17, 19, 83, 31, 39, 52, 63.

13. Schivelbusch, *Tastes of Paradise*; Sidney W. Mintz, "The Changing Roles of Food in the Study of Consumption," in *Consumption and the World of Goods*, ed. John Brewer (New York: Routledge, 1994), 261–73; Hole, *English Housewife*, 61.

14. Richard L. Bushman, *The Refinement of America: Persons, Houses, Cities* (New York: Vintage, 1993), xii.

15. Gibbs, "Taverns"; Wainwright, "Inns and Outs"; Robert Wormeley Carter, "Diary," 8 Jan. 1781, TS, Rockefeller Library

16. Ferdinard Marie Bayard, *Travels of a Frenchman in Maryland and Virginia*, ed. Ben C. McCary (Williamsburg, VA: Edwards Brothers, 1950), 4; Robert Honeyman, *Colonial Panorama 1775*, ed. Philip Padelford (San Marino, CA: Huntington Library, 1939); Alexander Macaulay, "Journal," WMQ 11, no. 3 (Jan. 1903) 184. For women roasting coffee beans, see *Martha Ogle Forman, Plantation Life at Rose Hill: The Diaries of Martha Ogle Forman 1814–1845*, ed. W. Emerson Wilson (Wilmington: Historical Society of Delaware, 1976) 30 Oct. 1829, 15 Jan. 1830; Harry Toulmin, *Western Country in 1793*, 35.

17. Luicinda Orr, *Journal of a Young Lady of Virginia*, 1782, 19 Sept. 1782 (Baltimore, 1871); Frances Baylor Hill, "The Diary of Frances Baylor Hill of Hillsborough, King and Queen County, Virginia (1797)," ed. William K. Bottorff and Roy Flannagan, 30 March 1797, *Early American Literature Newsletter* 2 (1967).

18. "Extracts from the Journal of Mrs. Ann Manigault, 1754–1781," Dec. 1754, *North American Women's Letters and Diaries*; Letter from Margaret Bayard Smith, Nov. 16, 1800, in *The First Forty Years of Washington Society in the Family Letters of Margaret Bayard Smith*, ed. Gaillard Hunt (New York: Frederick Ungar Publishing, 1906), 424; "Letter from Eleanor Parke Custis Lewis to Elizabeth Bordley, March 20, 1798," in *George Washington's Beautiful Nelly: The Letters of Eleanor Parke Custis*

Lewis to Elizabeth Bordley Gibson, 1794–1851, ed. Patricia Brady (Columbia: University of South Carolina Press, 1991), 287; Letter from Susanna Stuart Fitzhugh Knox, April 20, 1799, in "Reminiscences of the Knox and Soutter families of Virginia," *North American Women's Letters and Diaries*.

19. William Byrd, *Another Secret Diary of William Byrd of Westover 1739–1741*, Maude H. Woodfin and Marion Tingling, eds. (Richmond, VA: Dietz Press, 1942), 15,18; *The Journal of John Harrower: An Indentured Servant in the Colony of Virginia, 1773–1776*, ed. Edward Miles Riley (New York: Holt, Rinehart and Winston for Colonial Williamsburg, 1963), 56; "Olney Winsor Letters, 1786–1788," 7 Sept. 1786, microfilm, LVA; *Moreau de St. Mery's American Journey [1793–1798]*, ed. Kenneth Roberts and Anna M. Roberts (New York: Doubleday, 1947), 6; *Martha Ogle Forman, Plantation Life at Rose Hill: The Diaries of Martha Ogle Forman 1814–1845*, ed. W. Emerson Wilson, 6 Nov. 1816 (Wilmington: Historical Society of Delaware, 1976); "Letter from Sarah Cary to Polly Gray, January 01, 1786," in *The Cary Letters.*, ed. Caroline Curtis (Cambridge, MA: Riverside Press, 1891), 335.

20. Sarah Wister, "Diary of Sally Wister, December, 1777," in *Sally Wister's Journal: A True Narrative Being a Quaker Maiden's Account of Her Experiences with Officers of the Continental Army, 1777–1778*, ed. Albert Cook Myers (Philadelphia: Ferris and Leach Publishers, 1902), 224; Sarah Scofield Frost, "Diary of Sarah Scofield Frost, May, 1783," in *North American Women's Letters and Diaries*, 5.

21. *Minute Book Abstracts of Fauquier County, Virginia 1763–1764*, ed. Ruth and Sam Sparacio, 30 June 1764, (McLean, VA: Antient Press, 1994), 4:66; Thomas Griffin Peachy, "Memorandum Book, 1796–1810," 23 Sept. 1796, Photostat, Rockefeller Library; Walsh et al., *Provisioning Early American Towns*, 110; Bushman, *Refinement of America*, 184; Niemcewicz, *Under Their Vine and Fig Tree*, 80–81, 100.

22. Benjamin Franklin, *Poor Richard, An Almanack* (1738; New York: David McKay Company, 1976), 61; Theophilus Wreg, *The Virginia Almanack for the Year of our Lord God 1766* (Williamsburg, 1765), n.p., LCP; Philo Copernicus (pseud.), *The American Calendar; or, an Almanack, for the Year of our Lord 1771* (Philadelphia, 1770), n.p., LCP; *Virginia Gazette*, 14 Sept. 1739; *Pennsylvania Gazette*, 1 Feb. 1783.

23. Niemcewicz, *Under Their Vine and Fig Tree*, 119; Ellen G. D'Oench, *Prodigal Son Narratives 1480–1980* (New Haven: Yale University Art Gallery, 1995), 13, 11.

24. Francis Jerdone to Capt. William Thomson, "Letterbook of Francis Jerdone," 22 July 1752, 236, MS, LVA; Joshua Johnson to the firm, 17 April 1772, *Joshua Johnson's Letterbook, 1771–1774: Letters from a Merchant in London to his Partners in Maryland*, ed. Jacob Price, (London: London Record Society), 32; *Virginia Gazette*, 28 March 1751, 7 Jan. 1768, 10 March 1768; *Alexandria Gazette*, 3 Aug. 1803; William Fitzhugh to John Cooper, 20 Aug. 1690, and William Fitzhugh to George Luke, 20 May 1691, in *William Fitzhugh and His Chesapeake World, 1676–1701*, ed. Richard Beale Davis (Chapel Hill: Published for the Virginia Historical Society by the University of North Carolina Press, 1963), 268, 275; Browne, 1 March 1755; Charlotte Browne, "Diary," in "With Braddock's Army: Mrs. Browne's Diary in Virginia and Maryland," ed. Fairfax Harrison, *VMHB* 32, no. 4 (Oct. 1924), 1 March 1755.

25. Col. Francis Taylor, "Diary," 7 April 1792; Robert Carter to Messrs. James

Buchanan & Co., "Letterbook of Robert Carter III," 1 July 1761, TS, Rockefeller Library; C. A. Neal, *The Temperance Cook Book* (Philadelphia, 1843), 92–93, Library Company of Philadelphia.

26. Devereux Jarratt, *The Life of the Reverend Devereux Jarratt, Rector of Bath Parish, Dinwiddie County, Virginia, Written by Himself, in a Series of Letters Addressed to the Rev. John Coleman* (Baltimore, 1806), 13–14, Alderman Library.

27. *Virginia Gazette*, 10 Dec. 1767.

28. *Virginia Gazette*, 7 Jan. 1768, 4 Feb. 1768, 20 April 1769; *Maryland Gazette*, 29 Nov. 1764; *Virginia Gazette*, 7 Sept. 1769.

29. *Virginia Gazette*, 30 Aug. 1770, 27 Sept. 1770; *Maryland Gazette*, 2 Dec. 1762, 24 May 1759; *Virginia Gazette*, 7 Sept. 1769.

30. Landon Carter, "Diary," 2:1001; Byrd, *Secret Diary*, 459; *Virginia Gazette*, 4 May 1769, 19 Oct. 1769; *Maryland Gazette*, 29 Nov. 1764, 16 May 1765, 17 June 1773; *Virginia Gazette*, 14 June 1770.

31. Byrd, *Another Secret Diary*, 92, 129.

32. "Narrative of James Curry," *Slave Testimony: Two Centuries of Letters, Speeches, Interviews, and Autobiographies*, ed. John W. Blassingame (Baton Rouge: Louisiana State University, 1977), 137, 128; "Lewis Clarke," in Blassingame, *Slave Testimony*, 162.

33. "Narrative of James Fisher," in Blassingame, *Slave Testimony*, 231; "Interview of James Madison by Henry Bibb, 1851," ibid., 268; "Interview of James Smith by Henry Bibb, 1852," ibid., 281.

34. "Interview of Levi Douglass by James Wright, 1853," ibid., 304; "Sella Martin, 1867," ibid., 724; Forman, "Diary," 24 July 1817.

35. James Kirke Paulding to Morris Smith Miller, 27 Feb. 1813, *James Kirke Paulding, Forgotten Letter Writer*, ed. Ralph M. Aderman (Milwaukee, WI: n.p., 1957); William Fitzhugh to Richard Page, 8 Aug. 1690, 10 July 1683, in *William Fitzhugh*, 275, 148.

36. *Records of the Tuesday Club of Annapolis 1745–1756*, ed. Elaine G. Breslaw, 21 May 1745, 4 June 1745, 16 July 1745, 5 Aug. 1746, 11 Nov. 1746, 7 Nov. 1747, 10 Nov. 1747, 16 May 1749 (Chicago: University of Illinois Press, 1988).

37. *Virginia Gazette*, 16 June 1768; A. Cunyngham, *Virginia Gazette*, 21 July 1768; Anonymous, *Virginia Gazette*, 9 Feb. 1769; Benjamin Franklin, *Poor Richard, An Almanack* (1737; New York: David McKay Company, 1976), 60.

CONCLUSION

1. Orr, *Journal of a Young Lady*, 22 Sept. 1782.

Essay on Sources

Visitors to early Virginia sometimes kept journals on their excursions, and the resulting travel literature provides a valuable glimpse of the customs of the region. The most useful accounts include: Durand de Dauphine, *A Huguenot Exile in Virginia: or Voyages of a Frenchman Exiled for his Religion with a Description of Virginia and Maryland,* ed. Gilbert Chinard (1687; New York: Press of the Pioneers, 1934); Francis Louis Michel, "Report of the Journey of Francis Louis Michel from Berne, Switzerland, to Virginia October 2, 1701 — December 1, 1702," trans. and ed. William J. Hinke, *VMHB* 24, no. 2 (1916); Peter Kalm, *Travels in North America, 1748–1749,* trans. John Reinhold Forester (Barre, MA: Imprint Society, 1972); William Mylne, *Travels in the Colonies, 1773–1775: Described in the Letters of William Mylne,* ed. Ted Ruddock (Athens: University of Georgia Press, 1993); Nicholas Cresswell, *The Journal of Nicholas Cresswell, 1774–1777,* ed. Lincoln MacVeagh (New York: Dial Press, 1924); Philip Vickers Fithian, *Journal 1775–1776: Written on the Virginia-Pennsylvania Frontier and in the Army around New York,* ed. Robert Albion and Leonidas Dodson (Princeton: Princeton University Press, 1934); Ebenezer Hazard, "The Journal of Ebenezer Hazard in Virginia, 1777," ed. Fred Shelley, *VMHB* 62, no. 4 (1954); Francois Jean Marquis de Chastellux, *Travels in North America in the Years 1780, 1781, and 1782: A Revised Translation,* ed. Howard C. Rice Jr. (Chapel Hill: Published for the IEAHC by University of North Carolina Press, 1963); Francisco de Miranda, *The New Democracy in America: Travels of Francisco de Miranda in the United States, 1783–1784,* ed. John S. Ezell, trans. Judson P. Wood (Norman: University of Oklahoma Press, 1963); Ferdinand Marie Bayard, *Travels of a Frenchman in Maryland and Virginia,* ed. and trans. Ben C. McCary (1791; Williamsburg, VA: Edwards Brothers, 1950); Moreau de St. Mery, *American Journey [1793–1798],* trans. and ed. Kenneth Roberts and Anna M. Roberts (Garden City, NY: Doubleday, 1947); Lucious Verus Bierce, *Travels in the Southland 1822–1823: The Journal of Lucious Verus Bierce,* ed. George W. Knepper (Columbus: Ohio State University Press, 1966);

and J. P. Brissot de Warville, *New Travels in the United States of America: Performed in 1788*, ed. Durand Echeverria (1791; Cambridge: Belknap Press of Harvard University Press, 1964).

Most travel literature was written by men because men were freer to travel and because they enjoyed much higher rates of literacy than did women. In contrast, English housewifery literature (guides for women about how to manage the household) offers details about women's roles. Some of the most revealing include: Thomas Tusser, *Thomas Tusser 1557 Floruit: His Good Points of Husbandry*, ed. Dorothy Hartley (New York: Augustus M. Kelley, 1970); Gervase Markham, *Country Contentments, or The English Huswife* (London, 1623); George Savile, *The Lady's New-Year's Gift: or Advice to a Daughter* (London, 1688), microfilm, Alderman Library; Richard Bradley, *The Country Housewife and Lady's Director*, ed. Caroline Davidson (1727; London: Prospect Books, 1980); Eliza Smith, *The Compleat Housewife* (1727; London: Studio Editions, 1994); Eliza Haywood, *A Present for a Servant-Maid* (London, 1743); Martha Bradley, *The British Housewife* (London, 1770), microfilm, Alderman Library; Mary Cole, *The Lady's Complete Guide; or, Cookery in all its Branches* (1788; 2nd ed., London, 1791), microfilm, Alderman Library; Anon., *The Servants' Guide and Family Manual* (1830; 2nd ed., London, 1831); Robert Roberts, *The House-Servant's Directory: or, a Monitor for Private Families* (1827; 2nd ed., Boston, 1828), microfilm, Alderman Library; and Susannah Carter, *The Frugal Colonial Housewife*, ed. Jean McKibbin (1765; Garden City, NY: Dolphin Books, 1976).

Cookbooks are underutilized sources that offer another peek into women's lives. See Kenelme Digbie, *The Closet of the Eminently Learned Sir Kenelme Digbie Kt. Opened*, ed. Jane Stevenson and Peter Davidson (1669; Boston: Prospect Books, 1997); Harriott Pinckney Horry, *A Colonial Plantation Cookbook: The Receipt Book of Harriott Pinckney Horry*, ed. Richard J. Hooker (1770; Columbia: University of South Carolina Press, 1984); Martha Washington, *Martha Washington's Booke of Cookery and Booke of Sweetmeats: Being a Family Manuscript*, trans. Karen Hess (1749; New York: Columbia University Press, 1981); Amelia Simmons, *The First American Cook Book: A Facsimile of "American Cookery,"* ed. Mary Tolford Wilson (1796; New York: Dover, 1958); Mary Randolph, *The Virginia Housewife; or, Methodical Cook* (Philadelphia, 1846); and C. A. Neal, *The Temperance Cookbook* (3rd ed., Philadelphia, 1843), LCP.

Planter journals are invaluable sources of information regarding customs, labor practices, and consumption on plantations. The most thorough journals for large plantations are William Fitzhugh, *William Fitzhugh and His Chesapeake World 1676–1701: The Fitzhugh Letters and Other Documents*, ed. Richard Beale

Davis (Chapel Hill: Published for the Virginia Historical Society by University of North Carolina Press, 1963); William Byrd, *The Secret Diary of William Byrd of Westover, 1709–1712*, ed. Louis B. Wright and Marion Tinling (Richmond, VA: Dietz Press, 1941); *Another Secret Diary of William Byrd of Westover, 1739–1741*, ed. Maude H. Woodfin (Richmond, VA: Dietz Press, 1942); *The Diary of Colonel Landon Carter of Sabine Hall, 1752–1778*, ed. Jack P. Greene (Charlottesville: Published for the Virginia Historical Society by the University Press of Virginia, 1965); Robert Wormeley Carter, "Diary 1765–1792," TS, Rockefeller Library; and *The Papers of George Washington: Retirement Series*, ed. Dorothy Twohig and W. W. Abott (Charlottesville: University Press of Virginia, 1998-). In addition, *Records of the Tuesday Club of Annapolis, 1745–1756*, ed. Elaine G. Breslaw (Chicago: University of Illinois Press, 1988) offers a portrait of upper sort social life.

Journals and observations on middling plantations include: William Hugh Grove, "Diary 1698–1732,"MS, Alderman Library; John Fontaine, *The Journal of John Fontaine, An Irish Huguenot Son in Spain and Virginia 1710–1719*, ed. Edward Porter Alexander (Williamsburg: Colonial Williamsburg Foundation, 1972); Charlotte Browne, *Diary, 17 Nov. 1754 – 19 Jan. 1757*, photostat, VHS; Colonel James Gordon, "Journal of Col. James Gordon of Lancaster County (VA), 1758–1768," MS, VHS; John Harrower, *The Journal of John Harrower: An Indentured Servant in the Colony of Virginia, 1773–1776*, ed. Edward Miles Riley (New York: Published by Holt, Rinehart and Winston for Colonial Williamsburg, 1963); Sarah Fouace Nourse, "Diary 1781–1783," TS, Rockefeller Library; *Journal of Alexander Macaulay*, WMQ 11, no. 3; Colonel Francis Taylor, "Diary 1786–1799," microfilm, LVA; and the *Room-by-Room Inventories 1646–1824*, MS, Rockefeller Library.

Newspapers and almanacs were the least expensive printed materials in the Chesapeake and were circulated among family and friends. Colonists who could not read could hear them read aloud in the taverns and in homes. Newspapers and almanacs can thus be read, with caution, to determine what the lower sort knew. *The Virginia Gazette* (Williamsburg, 1736–1780), *The Maryland Gazette* (Annapolis, 1745–1839), and various Virginia and Maryland almanacs reveal the spread of distilling knowledge and its masculinization.

Court records offer another glimpse of the experiences of the lower sort who could not write and so did not leave much in the way of historical records. In recent years several individuals have published abstracts of hundreds of court proceedings. In particular, Joann Riley McKey compiled order abstracts for Accomack County, Virginia, 1663–1724 (Bowie, MD: Heritage Books, 1996–2001:), 14 vols.; Gibson Jefferson McConnaughey compiled order abstracts for Amelia County, Virginia, 1732–1760 (Amelia, A: Mid-South Publishing Company, 1985);

John Frederick Dorman compiled order abstracts for Caroline County, Virginia, 1732–1760 (Washington, DC, 1965–1982), 15 vols., and for Westmoreland County 1690–1721 (Washington, DC: 1962–1990), 15 vols.; June Whitehurst Johnson compiled the will books of Fairfax County, Virginia, from 1742–1752 and 1752–1767 (Berryville, VA: Virginia Book Co., 1982); Ruth and Sam Sparacio compiled the orders book abstracts for Lancaster County, Virginia, from 1656–1696 (McLean, VA: Antient Press, c.1993–c.1995), 8 vols., for Northumberland County, Virginia, 1661–1665 (McLean, VA: Antient Press, c.1994–c.1995), 5 vols., and for Stafford County, Virginia, 1664–1693 (n.p., 1987–1988), 2 vols.; Robert K. Headley Jr., compiled abstracts of wills for Richmond County, Virginia, 1699–1800 (Baltimore: Clearfield Company, 1993); Eliza Timberlake Davis abstracted the wills and administrations of Surry County, Virginia, from 1652 to 1750 (Baltimore: Clearfield Company, 1980), 3 vols.; Weynette Parks Haun compiled the Surry County, Virginia, court records from 1652 thru 1751, (Durham, NC: n.p., c.1987–c.1997), 10 vols.; Lindsay O. Duvall abstracted wills, deeds, and orders from several counties from 1678 to 1719 in *Colonial Virginia Abstracts* (Easley, SC: Southern Historical Press, c.1978–1979), 6 vols. Other useful court records include: Charles City County, Virginia, Order Book Aug. 1722 to Feb. 1722/3, MS, VHS; *Lower Norfolk County, Virginia, Court Records: Book "A" 1637 to 1746 and Book "B" 1646–1652*, transcribed by Alice Granbery Walter (Baltimore: Clearfield Company, 1994); Northampton Order Book 4, October 1658 to June 28, 1661, transcribed by Susie May Ames, MS, VHS; Peter Charles Hoffer and William B. Scott, eds., *Criminal Proceedings in Colonial Virginia: [Records of] Fines, Examination of Criminals, Trials of Slaves, etc., from March 1710 [1711] to [1754] [Richmond County, Virginia]* (Athens: Published for the American Historical Association by the University of Georgia Press, 1984); and W. W. Henning, *The Statutes at Large: Being a Collection of all the Laws of Virginia from the First Session of the Legislature in the Year 1619* (Richmond, VA, 1809–23), 13 vols.

Letterbooks, collections of the copies of correspondence that colonists sometimes made, offer details about trade in the Chesapeake both with the Atlantic world and within neighborhoods. Particularly noteworthy are "John Custis Letterbook 1717–1744," TS, Rockefeller Library; *Letters of Robert Carter 1720–1727: The Commercial Interests of a Virginia Gentleman*, ed. Louis B. Wright (Westport, CT: Greenwood Press, 1970); *Brothers of the Spade: Correspondence of Peter Collinson of London, and of John Custis, of Williamsburg, Virginia, 1736–1746*, ed. E. G. Swem (Barre, MA: Barre Gazette, 1957); "Joseph Ball Letterbook 1743–1780," MS, LC; "John Baylor Letterbooks, 1749–1753; 1757–1765," photostat, VHS; "Richard Corbin Letterbook, 1758–1768," microfilm, LVA; Daniel Roberdeau, "Letterbook 1767–1791," MS, LC; *Joshua Johnson's Letterbook, 1771–1774: Letters from a Merchant in London to his Partners in Maryland*, ed. Jacob Price (London: London Record Society, 1979); and John Hook, "Letterbooks 1763–1784," MSS, LVA.

Import records and account books reveal macro and micro views of trade. For imports, the following provided significant sources of detail: "Ledger of Imports and Exports, Christmas 1724 to Christmas 1725" (PRO Customs 3/25), Virginia Colonial Records Project microfilm roll 210; "Ledger of Imports and Exports, Christmas 1734 to Christmas 1735" (PRO Customs 3/35), VCRP roll 210; "Ledger of Imports and Exports, Christmas 1744 to Christmas 1745" (PRO Customs 3/45), VCRP roll 210; "Ledger of Imports and Exports, Christmas 1754 to Christmas 1755" (PRO Customs 3/55), VCRP roll 232; "Ledger of Imports and Exports, Christmas 1764 to Christmas 1765" (PRO Customs 3/65), VCRP roll 233; "Ledger of Imports and Exports, Christmas 1774 to Christmas 1775" (PRO Customs 3/75), VCRP roll 233. Account books provide a view of trade through the experiences of one individual; see "John Mercer Account Book 1731–1767," MS, VHS; Carter Burwell, "Littletown Plantation Ledger, 1736–1746," microfilm, Rockefeller Library; [Charles Carter], "Ship Tavern Ledger 1752–1758,"MS, VHS; and James Barrett Southall, "Accounts 1768–1771," photostat, Rockefeller Library.

On the creation of scientific cidering and distilling, husbandry literature (advice to men on managing estates) and prescriptive literature for men on alcoholic beverage production are invaluable. The Dibner Library at the National American History Museum, the Library Company of Philadelphia, and Winterthur Library have strong collections of these materials, including the most significant and popular works: Conrad Gesner, *A New Book of Distillation of Waters*, trans. Peter Mowen (London, 1559), Dibner Library; Hugh Plat, *Divers Chimicall Conclusions Concerning the Art of Distillation* (London, 1594), Dibner Library; John Evelyn, *Sylva; or, a Discourse of Forest Trees* (London, 1664), Dibner Library; *The Philosophical Transactions of the Royal Society of London, from their Commencement, in 1665 to the year 1800; Abridged*, ed. Charles Hutton, George Shaw, and Richard Pearson (London, 1809), Dibner Library, 18 vols.; John Worlidge, *The Second Part of Vinetum Britannicum; or, a Treatise of Cider* (London, 1689), Winterthur Library; John Worlidge, *Mr. Worlidge's Two Treatises* (London, 1694), Dibner Library; Royal Dublin Society, *Instructions for Planting and Managing Hops and for Raising Hop-Poles* (Dublin, 1733), Winterthur Library; Ambrose Cooper, *The Complete Distiller* (London, 1757), Dibner Library; Thomas Chapman, *The Cyder-Maker's Instructor, Sweet-Makers Assistant, and Victualler's and Housekeeper's Director* (London, 1762), LCP; George Watkins, *The Compleat English*

Brewer (London, 1768), Winterthur Library; Alexander Morrice, *A Treatise on Brewing* (London,1802), Dibner Library; Harrison Hall, *Hall's Distiller* (Philadelphia, 1813), Dibner Library; William Coxe, *A View of the Cultivation of Fruit Trees, and the Management of Orchards and Cider* (Philadelphia, 1817), Dibner Library; John Tuck, *The Private Brewer's Guide to the Art of Brewing* (1817; London, 1822), Dibner Library; Thomas Thomson, *Brewing and Distillation* (Edinburgh, 1849), Dibner Library; and *The American Farmer*, ed. John S. Skinner (Baltimore, 1819–1825), 11 vols., VHS.

Family papers, individual papers, and descriptive accounts give additional details about alcohol on plantations, in neighborhoods, and during the American Revolution. Especially valuable texts include: *The Papers of Robert Morris*, ed. E. James Ferguson (Pittsburgh: University of Pittsburgh Press, 1975); *The Papers of George Washington*, ed. Dorothy Twohig (Charlottesville: University Press of Virginia, 1998-); "The Papers of Joseph Trumbull 1753–1791," microfiche, David Library; "The Carter Papers," MSS, Alderman Library; "The Roberdeau Papers," MSS, LC; "The Mercer Papers," MSS, VHS; "The Jerdone Papers 1762–1801," MSS, Swem Library; "The Jefferson Papers," MSS, Alderman Library; Hugh Jones, *The Present State of Virginia*, ed. Richard L. Morton (1724; Chapel Hill: Published for the Virginia Historical Society by University of North Carolina Press, 1956); St. John DeCrevecoeur, *Sketches of Eighteenth Century America*, ed. Henri L. Bourdin, Ralph H. Gabriel, and Stanley T. Williams (New Haven: Yale University Press, 1925); and Robert Beverley, *The History and Present State of Virginia*, ed. Louis B. Wright (Charlottesville, VA: Dominion Books, 1947).

European historians have published a variety of monographs on alcoholic beverage production and consumption. Some of the most significant works are Judith M. Bennet, *Ale, Beer, and Brewsters in England: Women's Work in a Changing World, 1300–1600* (Oxford: Oxford University Press, 1996); Thomas Brennan, *Public Drinking and Popular Culture in Eighteenth-Century Paris* (Princeton: Princeton University Press, 1988); Peter Clark, *The English Alehouse: A Social History, 1200–1830* (London: Longman, 1983); A. Lynn Martin, *Alcohol, Sex, and Gender in Late Medieval and Early Modern Europe* (New York: Palgrave, 2001); and Richard W. Unger, *A History of Brewing in Holland 900–1900: Economy, Technology and the State* (Leiden: Brill, 2001). There are also excellent works on alcoholic beverage production in Latin America and Africa, including Jane Erin Mangan, *Trading Roles: Gender, Ethnicity and the Urban Economy in Colonial Potosi* (Durham, NC: Duke University Press, 2005), and Emmanuel Akyeampong, *Drink, Power, and Cultural Change: A Social History of Alcohol in Ghana, c. 1800 to Recent Times* (Oxford: James Currey, 1996).

On alcohol consumption in early America the most important books are David W. Conroy, *In Public Houses: Drink and the Revolution of Authority in Colonial Massachusetts* (Chapel Hill: Published for the OIEAHC by University of North Carolina Press, 1995); Peter Mancall, *Deadly Medicine: Indians and Alcohol in Early America* (Ithaca, NY: Cornell University Press, 1995); W. J. Rorabaugh, *The Alcoholic Republic: An American Tradition* (Oxford: Oxford University Press, 1979); and Peter Thompson, *Rum Punch and Revolution: Taverngoing and Public Life in Eighteenth-Century Philadelphia* (Philadelphia: University of Pennsylvania Press, 1999). Important analyses on taverns in particular include: Patricia Gibbs, "Taverns in Tidewater Virginia, 1700–1774" (master's thesis, College of William and Mary, 1968); Kym Rice, *Early American Taverns: For the Entertainment of Friends and Strangers* (Chicago: Regnery Gateway, 1983); and Sharon V. Salinger, *Taverns and Drinking in Early America* (Baltimore: Johns Hopkins University Press, 2002). For general works on alcoholic beverages in American history, see Stanley Baron, *Brewed in America: A History of Beer and Ale in the United States* (Boston: Little, Brown, 1962); and Andrew Barr, *Drink: A Social History of America* (New York: Carroll and Graf, 1999).

Anyone desiring to learn more about colonial Virginia and the early Chesapeake would do well to begin with Edmund Morgan's classic, *American Slavery, American Freedom: The Ordeal of Colonial Virginia* (New York: W.W. Norton, 1975). Kathleen Brown added gender to Morgan's account in *Good Wives, Nasty Wenches, and Anxious Patriarchs: Gender, Race, and Power in Colonial Virginia* (Chapel Hill: published for the IEAHC by University of North Carolina Press, 1996). The research and publications of the "Maryland School" have challenged and further refined Morgan's analysis. Morgan argued that Virginians (and by extension all southerners) were lazy and depended on enslaving African Americans to build white freedoms. Maryland School scholars have argued convincingly that Virginians' apparent laziness was in reality a series of rational labor-saving decisions. See Lois Green Carr, Russell R. Menard, and Lorena S. Walsh, *Robert Cole's World: Agriculture and Society in Early Maryland* (Chapel Hill: published for the OIEAHC by University of North Carolina Press, 1991); and Lois Green Car, Philip D. Morgan, and Jean B. Russo, *Colonial Chesapeake Society* (Chapel Hill: published for the IEAHC by University Press of North Carolina, 1988). James Horn examined immigration patterns to the Chesapeake and how colonists acclimated to the new world in *Adapting to a New World: English Society in the Seventeenth-Century Chesapeake* (Chapel Hill: published for the IEAHC by University of North Carolina Press, 1994). Darrett B. and Anita H. Rutman's *A Place in Time: Middlesex County, Virginia 1650–1750* (New York: W.W. Norton, 1984) supports the conclusion that the majority of Chesapeake colonists were provincial and local. Finally, the Colonial Williamsburg Early American History Research Reports at the Rockefeller Library and on microcard at Alderman Library are a priceless source of detail about the material culture of early Virginia.

Those interested in women's and gender history will find several monographs on early white southern women, including Cynthia A. Kierner's excellent *Beyond the Household: Women's Place in the Early South, 1700–1835* (Ithaca, NY: Cornell University Press, 1998), which explores how gender trumped class in the early South. Terri L. Snyder, in *Brabbling Women: Disorderly Speech and the Law in Early Virginia* (Ithaca, NY: Cornell University Press, 2003), makes the case that women enjoyed clout through the policing power of gossip until the eighteenth century. Linda Sturtz, in *Within Her Power: Propertied Women in Colonial Virginia* (New York: Routledge, 2002), argued that early Virginian women who owned property were feminists who had power over men. My research revealed no incipient feminism or men fearful of colonial women's "power," but rather that men and women generally worked together to survive in the new world.

One of the great disappointments about writing this book was that there was not more information available to include about Africans and African Americans. Philip Morgan's *Slave Counterpoint: Black Culture in the Eighteenth-Century Chesapeake and Lowcounty* (Chapel Hill: Published for the OIEAHC by University of North Carolina Press, 1998) does indeed serve as a counterpoint to all the books, this one included, that do not include nearly enough about African American experiences in early America. What little has been included in the preceding chapters is owed to *Slave Counterpoint* and to Deborah Gray White's *"Ar'n't I a Woman?": Female Slaves in the Plantation South* (New York: W.W. Norton, 1985). While Charles Joyner's *Down by the Riverside: A South Carolina Slave Community* (Urbana: University of Chicago Press, 1984) clearly focuses on South Carolina, this powerful book can help one to imagine life and labor on Chesapeake plantations for both black and white.

Anyone interested in learning more about gender and early modern science should begin with two books by Londa Schiebinger: *The Mind Has No Sex? Women in the Origins of Modern Science* (Cambridge: Harvard University Press, 1989), and *Nature's Body: Gender in the Making of Modern Science* (Boston: Beacon, 1993), that elaborate on the impact of the founding of the Royal Society of Science and on why women were deliberately kept out of English science. Evelyn Fox Keller has expanded on these topics in *Reflections of Gender and Science* (New Haven: Yale University Press, 1985).

On technology and science in the New World, Joyce Chaplin's *An Anxious*

Pursuit: Agricultural Innovation and Modernity in the Lower South, 1730–1815 (Chapel Hill: published for the IEAHC by the University of North Carolina Press, 1993), is outstanding. Judith A. McGaw edited a collection of essays on *Early American Technology: Making and Doing Things from the Colonial Era to 1850* (Chapel Hill: published for the OIEAHC by the University of North Carolina Press, 1994), which demonstrates, as does Chaplin's book, that early Americans were deeply desirous of and invested in technological development. Raymond Phineas Stearns, *Science in the British Colonies of America* (Urbana: University of Illinois Press, 1970) details the spread of British science discoveries and philosophies to the New World.

Multiple works on supplying the revolutionary army are both useful and fascinating, although none discuss alcohol provisions specifically. See: E. Wayne Carp, *To Starve the Army at Pleasure: Continental Army Administration and American Political Culture 1775–1783* (Chapel Hill: University of North Carolina Press, 1984); Holly A. Mayer, *Belonging to the Army: Camp Followers and Community during the American Revolution* (Columbia: University of South Carolina Press, 1996); Erma Risch, *Quartermaster Support of the Army: A History of the Corps 1775–1939* (Washington, DC: Quartermaster Historian's Office, Office of the Quartermaster General, 1962) and *Supplying Washington's Army* (Washington, DC: Center of Military History, United States Army, 1981); and Charles Royster, *A Revolutionary People at War: The Continental Army and American Character, 1775–1783* (Chapel Hill: Published for the IEAHC by University of North Carolina Press, 1979).

Anyone interested in temperance should read W. J. Rorabaugh's *Alcoholic Republic*, which argues convincingly that temperance was a reaction to the spread of cheap western corn whiskey in the early nineteenth century. Most research on temperance has focused on New England or the West. Two exceptions are C. C. Pearson and J. Edwin Hendricks, *Liquor and Anti-Liquor in Virginia, 1619–1919* (Durham, NC: Duke University Press, 1967) and Joe L. Coker, *Liquor in the Land of the Lost Cause: Southern White Evangelicals and the Prohibition Movement* (Lexington: University Press of Kentucky, 2007). *Drinking: Behavior and Belief in Modern History*, edited by Susanna Barrows and Robin Room (Berkeley: University of California Press, 1991) is also a useful resource for the study of alcohol consumption and temperance. Those interested in gender and temperance should read Scott Martin's excellent *Devil of the Domestic Sphere: Temperance, Gender, and Middle-class Ideology, 1800–1860* (DeKalb: Northern Illinois University Press, 2008).

Index